Test Yourself

Business Calculus

Lawrence A. Trivieri, M.A.
DeKalb College
Clarkston, GA

Contributing Editors

Mark Weinfeld, M.A.
President
MATHWORKS
New York, NY

Shared Keny, Ph.D.
Department of Mathematics
Whittier College
Whittier, CA

Tony Julianelle, Ph.D.
Department of Mathematics
University of Vermont
Burlington, VT

NTC LearningWorks
a division of NTC Publishing Group
Lincolnwood, Illinois

Library of Congress Cataloging-in-Publication Data
is available from the Library of Congress.

ISBN 0844223522

A *Test Yourself Books, Inc.* Project

Published by NTC Publishing Group
© 1996 NTC Publishing Group, 4255 West Touhy Avenue
Lincolnwood (Chicago), Illinois 60646-1975 U.S.A.

6 7 8 9 ML 0 9 8 7 6 5 4 3 2 1

Contents

Preface

This book has been written so that you can test your knowledge of business calculus. It is not a textbook but is based upon the leading texts for such a course. Use this book in conjunction with your own textbook if you are currently enrolled in a business calculus course. It can also be used as a review book to refresh your knowledge of the subject material if you have been out of school for a while.

Each chapter begins with a set of sample questions for you to answer. Within this section, key concepts, terms, and equations appear in boxes, where you will find concise explanations, as well. After working through the exercises, you'll find complete solutions in the Check Yourself section. Every answer is explained thoroughly, including a reference to which specific subtopic you need to study if you missed that question. Finally, each chapter concludes with a unique self-diagnostic grading chart that helps you pinpoint the topics and subtopics you need to review further.

Study the particular chapters of this book as you cover the corresponding material in your course. If this book is used as intended, you should be able to meet with great success in your course.

I wish to thank Fred N. Grayson for encouraging me to write this book and for his careful and thorough guidance leading to the completion of the project. Further, I wish to thank Sherry Francis for her very careful reading of the original manuscript.

You have my very best wishes for a successful experience.

Lawrence A. Trivieri

How to Use this Book

This "Test Yourself" book is part of a unique series designed to help you improve your test scores on almost any type of examination you will face. Too often, you will study for a test—quiz, midterm, or final—and come away with a score that is lower than anticipated. Why? Because there is no way for you to really know how much you understand a topic until you've taken a test. The *purpose* of the test, after all, is to test your complete understanding of the material.

The "Test Yourself" series offers you a way to improve your scores and to actually test your knowledge at the time you use this book. Consider each chapter a diagnostic pretest in a specific topic. Answer the questions, check your answers, and then give yourself a grade. Then, and only then, will you know where your strengths and, more important, weaknesses are. Once these areas are identified, you can strategically focus your study on those topics that need additional work.

Each book in this series presents a specific subject in an organized manner, and although each "Test Yourself" chapter may not correspond to exactly the same chapter in your textbook, you should have little difficulty in locating the specific topic you are studying. Written by educators in the field, each book is designed to correspond, as much as possible, to the leading textbooks. This means that you can feel confident in using this book, and that regardless of your textbook, professor, or school, you will be much better prepared for anything you will encounter on your test.

Each chapter has four parts:

Brief Yourself. All chapters contain a brief overview of the topic that is intended to give you a more thorough understanding of the material with which you need to be familiar. Sometimes this information is presented at the beginning of the chapter, and sometimes it flows throughout the chapter, to review your understanding of various *units* within the chapter.

Test Yourself. Each chapter covers a specific topic corresponding to one that you will find in your textbook. Answer the questions, either on a separate page or directly in the book, if there is room.

Check Yourself. Check your answers. Every question is fully answered and explained. These answers will be the key to your increased understanding. If you answered the question incorrectly, read the explanations to *learn* and *understand* the material. You will note that at the end of every answer you will be referred to a specific subtopic within that chapter, so you can focus your studying and prepare more efficiently.

Grade Yourself. At the end of each chapter is a self-diagnostic key. By indicating on this form the numbers of those questions you answered incorrectly, you will have a clear picture of your weak areas.

There are no secrets to test success. Only good preparation can guarantee higher grades. By utilizing this "Test Yourself" book, you will have a better chance of improving your scores and understanding the subject more fully.

Functions

 Test Yourself

1.1 Introduction

Definitions: A **function** is a rule or correspondence that assigns to each input element exactly one output element. The set of all input elements for which the rule or correspondence applies is called the **domain** of the function. The set of all output elements is called the **range**.

For each function in Exercises 1-11, determine the domain.

1. $f(x) = 3x$

2. $h(x) = \dfrac{3x - 2}{3}$

3. $F(u) = \dfrac{3}{u}$

4. $G(u) = \dfrac{u}{u - 1}$

5. $f(t) = |\, t + 1 \,|$

6. $g(t) = \sqrt{t}$

7. $F(y) = 1 \div \sqrt{y}$

8. $G(y) = \dfrac{y - 3}{y^2 - 5y + 6}$

9. $H(y) = 3y^2 - 4y + 7$

10. $f(x) = \begin{cases} 2, \text{ if } x \le 3 \\ 4, \text{ if } x > 3 \end{cases}$

11. $h(x) = \begin{cases} x, \text{ if } x < -2 \\ 2x + 1, \text{ if } -2 \le x < 3 \\ x^2, \text{ if } x > 3 \end{cases}$

1.2 Algebra of Functions

Definition: Let f and g be two functions. The **sum** of f and g, denoted by f + g, is defined as follows:

$$(f + g)(x) = f(x) + g(x)$$

for all x both in the domain of f and in the domain of g.

In Exercises 12-15, form the indicated sum of the two functions given.

12. $f(x) = 2x - 3$ and $g(x) = x^2 + 1$; $f + g$

13. $h(y) = \dfrac{y}{y + 3}$ and $j(y) = |\, y - 3 \,|$; $h + j$

14. $H(y) = \sqrt{y + 2}$ and $J(y) = \dfrac{1}{y^2}$; $H + J$

15. $T(u) = \sqrt{1 - 3u}$ and $V(u) = 2u^2 - 1$; $T + V$

Definition: Let f and g be two functions. The **difference** of f and g, denoted by f – g, is defined as follows:

$$(f - g)(x) = f(x) - g(x)$$

for all x both in the domain of f and in the domain of g.

In Exercises 16-19, form the indicated difference of the two functions given.

16. $f(x) = 2x^3 - 1$ and $g(x) = 4 - 5x$; $f - g$

17. $F(x) = \sqrt{2x - 1}$ and $G(x) = \sqrt{x + 3}$; $G - F$

18. $h(u) = \dfrac{1}{u}$ and $j(u) = |u + 3|$; $h - j$

19. $R(s) = |2s - 3|$ and $T(s) = \dfrac{2}{s - 3}$; $R - T$

Definition: Let f and g be two functions. The **product** of f and g, denoted by f.g, is defined as follows:

$$(f \cdot g)(x) = f(x) \cdot g(x)$$

for all x both in the domain of f and in the domain of g.

In Exercises 20-23, form the indicated product of the two functions given.

20. $f(u) = 3 - 5u$ and $g(u) = \dfrac{1}{u + 1}$; $f \cdot g$

21. $F(u) = \sqrt{u - 3}$ and $G(u) = \dfrac{1}{u^2 - 1}$; $G \cdot F$

22. $R(y) = \sqrt[3]{y - 1}$ and $S(y) = \sqrt{y - 7}$; $R \cdot S$

23. $S(x) = |x - 3|$ and $T(x) = \sqrt{2x - 5}$; $S \cdot T$

Definition: Let f and g be two functions. The quotient of f and g, denoted by $\dfrac{f}{g}$, is defined as follows:

$$\left(\frac{f}{g}\right)(x) = \frac{f(x)}{g(x)}$$

for all x both in the domain of f and in the domain of g, and such that $g(x) \neq 0$.

In Exercises 24-27, form the indicated quotient of the two functions given.

24. $f(x) = 3x - 2$ and $g(x) = x - 3$; $\dfrac{f}{g}$

25. $F(x) = \sqrt{2x - 3}$ and $G(x) = \dfrac{1}{x}$; $\dfrac{G}{F}$

26. $H(u) = |2u + 3|$ and $J(u) = u^2 - 3$; $\dfrac{H}{J}$

27. $T(y) = \sqrt{y + 4}$ and $V(y) = \dfrac{3}{y - 2}$; $\dfrac{T}{V}$

1.3 Composition of Functions

Definition: The **composite function** f of g, denoted by f o g, is defined as follows:

$$(f \circ g)(x) = f(g(x))$$

such that x is an element in the domain of g and g(x) is an element in the domain of f.

In Exercises 28-31, form the indicated composite functions.

28. $f(x) = x^3$ and $g(x) = 2x - 1$; $(f \circ g)$

29. $F(x) = \sqrt{x + 1}$ and $G(x) = 3x$; $(G \circ F)$

30. $C(r) = 2r - 3$ and $D(r) = |r - 1|$; $(C \circ D)$

31. $T(y) = \sqrt{3 - y}$ and $W(y) = y^2$; $(T \circ W)$

1.4 Inverse Functions

Definition: If f is a function such that each element in its range is used exactly once, then f is said to be a **one-to-one** function.

Definition: If f is a one-to-one function defined by $y = f(x)$, then the **inverse function** of f, denoted by f^{-1}, is a function such that

$f^{-1}(f(x)) = x$ for every x in the domain of f, and

$f(f^{-1}(x)) = x$ for every x in the domain of f^{-1}.

Procedure: To determine the inverse function, f^{-1}, of the function f, defined by $y = f(x)$:

Step 1: Write $y = f(x)$.

Step 2: Interchange the variables x and y.

Step 3: Solve for y in the equation formed in Step 2.

Step 4: If the equation formed in Step 3 defines a

function, write $y = f^{-1}(x)$. [The inverse function, $f^{-1}(x)$, does not always exists. Hence, it is important to check the equation in Step 3.]

In Exercises 32–35, determine the inverse function of the given function, if the inverse function exists. Also, determine the domain of the inverse function.

32. $f(x) = 2x + 5$

33. $g(x) = \sqrt{x + 3}$

34. $F(x) = \dfrac{2x - 3}{5}$

35. $G(x) = |x - 2|$

1.5 Graphs of Functions

Functions can be graphed in a coordinate plane. If f is a function, in the variable x, then the **graph** of f is the set of all points $(x, f(x))$ in the plane such that x is in the domain of f. Since each element in the domain of a function is paired with exactly one element in its range, any vertical line in the plane, intersecting the graph of a function, will intersect the graph exactly once. This is known as the **vertical line test** and is used to determine if a graph is the graph of a function.

In Exercises 36-40, use the vertical line test to determine whether the given figure is or is not the graph of a function.

36. A horizontal line.

37. An oblique line.

38. An ellipse.

39. The letter Q.

40. The letter W.

In Exercises 41-47, graph each of the given functions in the xy-plane.

41. $f(x) = \sqrt{x}$

42. $g(x) = 2x + 3$

43. $F(x) = |x + 3|$

44. $G(x) = \dfrac{1}{x}$

45. $P(x) = \sqrt{2x - 3}$

46. $H(x) = \begin{cases} 3, & \text{if } x > 2 \\ -2, & \text{if } x \le 2 \end{cases}$

47. $V(x) = \begin{cases} 2x - 3, & \text{if } x < 0 \\ x^2 + 3, & \text{if } x \ge 0 \end{cases}$

1.6 Linear Functions

Definition: The function L, in the variable x, is a **linear function** if and only if it can be written in the form $L(x) = ax + b$ such that a and b are real numbers and $a \ne 0$.

In Exercises 48-52, determine which of the given functions are linear functions.

48. $f(x) = 2x + 7$

49. $g(x) = \dfrac{5x - 1}{2}$

50. $F(x) = x^2 + 1$

51. $G(x) = -5x$

52. $T(x) = 1.2x - 3.7$

The graph of a linear function is an oblique line.

In Exercises 53-55, graph each of the following functions in the xy-plane.

53. $f(x) = x + 2$

54. $h(x) = \dfrac{2x + 1}{3}$

55. $F(x) = -2x + 3$

If the graph of the linear function $L(x) = ax + b$ $(a \ne 0)$ is in the xy–plane, then a is the slope of the line and *b* is its y-intercept.

In Exercises 56–58, the graph of each function is an oblique line. Determine its slope and y-intercept.

56. $y = f(x) = 2x - 5$

57. $y = h(x) = \dfrac{3x - 1}{4}$

58. $y = G(x) = -7x$

1.7 Quadratic Functions

Definition: The function f, in the variable x, is a **quadratic function** if and only if it can be written in the form $f(x) = ax^2 + bx + c$ such that a, b, and c are real numbers and $a \neq 0$.

In Exercises 59-63, determine which of the given functions are quadratic functions.

59. $f(x) = 3x^2 - 5x + 6$

60. $h(x) = 7 - 9x^2$

61. $F(x) = 8 - 9x$

62. $H(x) = 2x - 6x^2$

63. $J(x) = \sqrt{x^2 - 5x + 6}$

The graph of a quadratic function is a parabola with a vertical axis.

In Exercises 64-67, graph each of the following quadratic functions in the xy-plane.

64. $f(x) = x^2 + 2$

65. $h(x) = -2x^2 + 3x - 1$

66. $j(x) = 2x - x^2$

67. $F(x) = 3x^2 + 2x + 1$

✓ Check Yourself

1. The function, $f(x) = 3x$, is a polynomial function. Its domain is the set of all real numbers. In interval notation, the domain is $(-\infty, +\infty)$. (**Domain of a function**)

2. The function, $h(x) = \dfrac{3x - 2}{3}$, is a polynomial function. Its domain is $(-\infty, +\infty)$. (**Domain of a function**)

3. The function, $F(u) = \dfrac{3}{u}$, is a rational function. Its domain is the set of all real numbers such that the denominator, u, is not equal to 0. Hence, the domain is the set of all nonzero real numbers. In interval notation, the domain is $(-\infty, 0) \cup (0, +\infty)$. (**Domain of a function**)

4. The function, $G(u) = \dfrac{u}{u - 1}$, is a rational function. Its domain is the set of all real numbers such that the denominator, u – 1, is not equal to 0. (u – 1 = 0 if u = 1.) Hence, the domain is the set of all real numbers other than 1. In interval notation, the domain is $(-\infty, 1) \cup (1, +\infty)$. (**Domain of a function**)

5. The function, $f(t) = |t + 1|$, is a variation of the absolute value function. Its domain is the set of all real numbers, or $(-\infty, +\infty)$. (**Domain of a function**)

6. The function, $g(t) = \sqrt{t}$, is the square root function. Its domain is the set of all nonnegative real numbers (i.e., $t \geq 0$). Hence, its domain is $[0, +\infty)$. (**Domain of a function**)

7. The function, $F(y) = 1 \div \sqrt{y}$, is the reciprocal of the square root function. Its domain is the set of all positive real numbers, or $(0, +\infty)$. (**Domain of a function**)

8. The function, $G(y) = \dfrac{y-3}{y^2 - 5y + 6}$, is a rational function. Its domain is the set of all real numbers such that the denominator, $y^2 - 5y + 6$, is not equal to 0. ($y^2 - 5y + 6 = 0$ if $y = 2, 3$.) Hence, the domain is $(-\infty, 2)$ $\cup (2, 3) \cup (3, +\infty)$. **(Domain of a function)**

9. The function, $H(y) = 3y^2 - 4y + 7$, is a polynomial function. Its domain is the set of all real numbers, or $(-\infty, +\infty)$. **(Domain of a function)**

10. The function, $f(x) = \begin{cases} 2, \text{ if } x \le 3 \\ 4, \text{ if } x > 3 \end{cases}$ is a piecewise or multi-part function. Note that $f(x)$ is defined for all real values of x. If $x \le 3$, then $f(x) = 2$. If $f(x) > 3$, then $f(x) = 4$. Hence, the domain is $(-\infty, +\infty)$. **(Domain of a function)**

11. The function, $h(x) = \begin{cases} x, \text{ if } x < -2 \\ 2x + 1, \text{ if } -2 \le x < 3 \\ x^2, \text{ if } x > 3 \end{cases}$ is a piecewise or multi-part function. Note that $h(x)$ is defined for all real values of x other than 3. Hence, the domain is $(-\infty, 3) \cup (3, +\infty)$. **(Domain of a function)**

12. $f(x) = 2x - 3$ and $g(x) = x^2 + 1$

    $(f + g)(x) = f(x) + g(x)$

    $\qquad = (2x - 3) + (x^2 + 1)$

    $\qquad = 2x - 3 + x^2 + 1$

    $(f + g)(x) = x^2 + 2x - 2$

    Since the domain of f is $(-\infty, +\infty)$ and the domain of g is $(-\infty, +\infty)$, then the domain of $f + g$ is also $(-\infty, +\infty)$. **(Sum of functions)**

13. $h(y) = \dfrac{y}{y+3}$ and $j(y) = |\, y - 3\,|$

    $(h + j)(y) = h(y) + j(y)$

    $(h + j)(y) = \dfrac{y}{y+3} + |\, y - 3\,|$

    The domain of h is the set of all real numbers other than -3 and the domain of j is the set of all real numbers. Hence, the domain of $h + j$ is $(-\infty, -3) \cup (-3, +\infty)$. **(Sum of functions)**

14. $H(y) = \sqrt{y+2}$ and $J(y) = \dfrac{1}{y^2}$

    $(H + J)(y) = H(y) + J(y)$

    $(H + J)(y) = \sqrt{y+2} + \dfrac{1}{y^2}$

    The domain of H is $[-2, +\infty)$ and the domain of J is the set of all nonzero real numbers. Hence, the domain of $H + J$ is $[-2, 0) \cup (0, +\infty)$. **(Sum of functions)**

15. $T(u) = \sqrt{1 - 3u}$ and $V(u) = 2u^2 - 1$

    $(T + V)(u) = T(u) + V(u)$

    $\qquad = \sqrt{1 - 3u} + (2u^2 - 1)$

    $(T + V)(u) = \sqrt{1 - 3u} + 2u^2 - 1$

The domain of T is $(-\infty, \frac{1}{3}]$, and the domain of V is $(-\infty, +\infty)$.

Hence, the domain of T + V is $(-\infty, \frac{1}{3}]$. **(Sum of functions)**

16. $f(x) = 2x^3 - 1$ and $g(x) = 4 - 5x$

$(f - g)(x) = f(x) - g(x)$

$$= (2x^3 - 1) - (4 - 5x)$$

$$= 2x^3 - 1 - 4 + 5x$$

$(f - g)(x) = 2x^3 + 5x - 5$

The domain of f is $(-\infty, +\infty)$ and the domain of g is $(-\infty, +\infty)$. Hence, the domain of f – g is $(-\infty, +\infty)$. **(Difference of functions)**

17. $F(x) = \sqrt{2x - 1}$ and $G(x) = \sqrt{x + 3}$

$(G - F)(x) = G(x) - F(x)$

$(G - F)(x) = \sqrt{x + 3} - \sqrt{2x - 1}$

The domain of G is $[-3, +\infty)$ and the domain of F is $[\frac{1}{2}, +\infty)$.

Hence, the domain of G – F is $[\frac{1}{2}, +\infty)$. **(Difference of functions)**

18. $h(u) = \frac{1}{u}$ and $j(u) = |u + 3|$

$(h - j)(u) = h(u) - j(u)$

$(h - j)(u) = \frac{1}{u} - |u + 3|$

The domain of h is the set of all nonzero real numbers and the domain of j is $(-\infty, +\infty)$. Hence, the domain of h – j is $(-\infty, 0) \cup (0, +\infty)$. **(Difference of functions)**

19. $R(s) = |2s - 3|$ and $T(s) = \frac{2}{s - 3}$

$(R - T)(s) = R(s) - T(s)$

$(R - T)(s) = |2s - 3| - \frac{2}{s - 3}$

The domain of R is $(-\infty, +\infty)$ and the domain of T is $(-\infty, 3) \cup (3, +\infty)$. Hence, the domain of R – T is $(-\infty, 3) \cup (3, +\infty)$. **(Difference of functions)**

20. $f(u) = 3 - 5u$ and $g(u) = \frac{1}{u + 1}$

$(f \cdot g)(u) = f(u) \cdot g(u)$

$$= (3 - 5u)(\frac{1}{u + 1})$$

$(f \cdot g)(u) = \frac{3 - 5u}{u + 1}$

The domain of f is $(-\infty, +\infty)$ and the domain of g is $(-\infty, -1) \cup (-1, +\infty)$. Hence, the domain of f·g is $(-\infty, -1) \cup (-1, +\infty)$. **(Product of functions)**

21. $F(u) = \sqrt{u-3}$ and $G(u) = \dfrac{1}{u^2-1}$

$(G·F)(u) = G(u)·F(u)$

$$= \dfrac{1}{u^2-1} \cdot \sqrt{u-3}$$

$(G·F)(u) = \dfrac{\sqrt{u-3}}{u^2-1}$

The domain of G is the set of all real numbers other than ±1 and the domain of F is $[3, +\infty)$. Therefore, the domain of G·F is $[3, +\infty)$. **(Product of functions)**

22. $R(y) = \sqrt[3]{y-1}$ and $S(y) = \sqrt{y-7}$

$(R·S)(y) = R(y)·S(y)$

$(R·S)(y) = (\sqrt[3]{y-1})(\sqrt{y-7})$

The domain of R is $(-\infty, +\infty)$ and the domain of S is $[7, +\infty)$. Hence, the domain of R·S is $[7, +\infty)$. **(Product of functions)**

23. $S(x) = |x-3|$ and $T(x) = \sqrt{2x-5}$

$(S·T)(x) = S(x)·T(x)$

$(S·T)(x) = |x-3| \cdot \sqrt{2x-5}$

The domain of S is $(-\infty, +\infty)$ and the domain of T is $[\dfrac{5}{2}, +\infty)$.

Hence, the domain of $S \cdot T$ is $[\dfrac{5}{2}, +\infty)$. **(Product of functions)**

24. $f(x) = 3x - 2$ and $g(x) = x - 3$

$\left(\dfrac{f}{g}\right)(x) = \dfrac{f(x)}{g(x)}, g(x) \neq 0$

$\left(\dfrac{f}{g}\right)(x) = \dfrac{3x-2}{x-3}$

The domain of f is $(-\infty, +\infty)$ and the domain of g is $(-\infty, +\infty)$. However, $g(x) = 0$ if $x = 3$. Hence, the domain of $\dfrac{f}{g}$ is $(-\infty, 3) \cup (3, +\infty)$ **(Quotient of functions)**

25. $F(x) = \sqrt{2x-3}$ and $G(x) = \dfrac{1}{x}$

$\left(\dfrac{G}{F}\right)(x) = \dfrac{G(x)}{F(x)}, F(x) \neq 0$

$$= \dfrac{1}{x} \div \sqrt{2x-3}$$

$\left(\dfrac{G}{F}\right)(x) = 1 \div (x\sqrt{2x-3}) = \dfrac{1}{x\sqrt{2x-3}}$

The domain of G is the set of all nonzero real numbers and the domain of F is $[\frac{3}{2}$, +∞). However, F(x) = 0 if $x = \frac{3}{2}$. Hence, the domain of $\frac{G}{F}$ is $(\frac{3}{2}$, +∞). **(Quotient of functions)**

26. H(u) = | 2u + 3 | and J(u) = $u^2 - 3$

$$\left(\frac{H}{J}\right)(u) = \frac{H(u)}{J(u)}, \; J(u) \neq 0$$

$$\left(\frac{H}{J}\right)(u) = \frac{|2u + 3|}{u^2 - 3}$$

The domain of H is (−∞, +∞) and the domain of J is (−∞, +∞). However, J(u) = 0 if $u = \pm\sqrt{3}$. Hence, the domain of $\frac{H}{J}$ is $(-\infty, -\sqrt{3}) \cup (-\sqrt{3}, +\sqrt{3}) \cup (+\sqrt{3}, +\infty)$. **(Quotient of functions)**

27. T(y) = $\sqrt{y + 4}$ and V(y) = $\frac{3}{y - 2}$

$$\left(\frac{T}{V}\right)(y) = \frac{T(y)}{V(y)}, \; V(y) \neq 0$$

$$= \frac{\sqrt{y + 4}}{\left(\frac{3}{y - 2}\right)}$$

$$\left(\frac{T}{V}\right)(y) = \frac{(y - 2)\sqrt{y + 4}}{3}$$

The domain of T is [−4, +∞) and the domain of V is the set of all real numbers other than 2. However, V(y) is not defined if y = 2. Hence, the domain of $\frac{T}{V}$ is [−4, 2) ∪ (2, +∞). **(Quotient of functions)**

28. f(x) = x^3 and g(x) = 2x − 1

(f ∘ g)(x) = f(g(x))

$$= f(2x − 1)$$

$$= (2x − 1)^3$$

The domain of f ∘ g = {x | x is in the domain of g and g(x) is in the domain of f}

$$= \{x \mid x \text{ is in } (-\infty, +\infty) \text{ and } (2x − 1) \text{ is in } (-\infty, +\infty)\}$$

Hence, the domain of f ∘ g is (−∞, + ∞). **(Composition of functions)**

29. F(x) = $\sqrt{x + 1}$ and G(x) = 3x

(G ∘ F)(x) = G(F(x))

$$= G(\sqrt{x + 1})$$

$$= 3\sqrt{x + 1}$$

The domain of G ∘ F = {x | x is in the domain of F and F(x) is in the domain of G}

$$= \{x \mid x \text{ is in } [-1, +\infty) \text{ and } \sqrt{x + 1} \text{ is in } (-\infty, +\infty)\}$$

Hence, the domain of G ∘ F is [−1, +∞). **(Composition of functions)**

30. $C(r) = 2r - 3$ and $D(r) = |r - 1|$

   $(D \circ C)(r) = D(C(r))$

$$= D(2r - 3)$$

$$= |(2r - 3) - 1|$$

   $(D \circ C)(r) = |2r - 4|$

   The domain of $D \circ C = \{r \,|\, r$ is in the domain of C and C(r) is in the domain of D$\}$

$$= \{r \,|\, r \text{ is in } (-\infty, +\infty) \text{ and } (2r - 3) \text{ is in } (-\infty, +\infty)\}$$

   Hence, the domain of $D \circ C$ is $(-\infty, +\infty)$. **(Composition of functions)**

31. $T(y) = \sqrt{3 - y}$ and $W(y) = y^2$

   $(T \circ W)(y) = T(W(y))$

$$= T(y^2)$$

$$= \sqrt{3 - y^2}$$

   The domain of $T \circ W = \{y \,|\, y$ is in the domain of W and W(y) is in the domain of T$\}$

$$= \{y \,|\, y \text{ is in } (-\infty, +\infty) \text{ and } y^2 \text{ is in } (-\infty, 3]\}$$

$$= \{y \,|\, y \text{ is in } (-\infty, +\infty) \text{ and } y \text{ is in } [-\sqrt{3}, \sqrt{3}]\}$$

   Hence, the domain of $T \circ W$ is $[-\sqrt{3}, \sqrt{3}]$. **(Composition of functions)**

32. $f(x) = 2x + 5$; x is in $(-\infty, +\infty)$, $f(x)$ is in $(-\infty, +\infty)$

   1. $y = 2x + 5$; x is in $(-\infty, +\infty)$, y is in $(-\infty, +\infty)$

   2. $x = 2y + 5$

   3. $2y = x - 5$

   $$y = \frac{x - 5}{2}$$

   4. $f^{-1}(x) = \dfrac{x - 5}{2}$, x is in $(-\infty, +\infty)$ **(Inverse of a function)**

33. $g(x) = \sqrt{x + 3}$; x is in $[-3, +\infty)$, $g(x)$ is in $[0, +\infty)$

   1. $y = \sqrt{x + 3}$; x is in $[-3, +\infty)$, y is in $[0, +\infty)$

   2. $x = \sqrt{y + 3}$; y is in $[-3, +\infty)$, x is in $[0, +\infty)$

   3. $x^2 = y + 3$

   $$y = x^2 - 3 \; ; \; x \geq 0$$

   4. $f^{-1}(x) = x^2 - 3$; $x \geq 0$ **(Inverse of a function)**

34. $F(x) = \dfrac{2x - 3}{5}$; x is in $(-\infty, +\infty)$, $F(x)$ is in $(-\infty, +\infty)$

   1. $y = \dfrac{2x - 3}{5}$; x is in $(-\infty, +\infty)$, y is in $(-\infty, +\infty)$

   2. $x = \dfrac{2y - 3}{5}$; y is in $(-\infty +\infty)$, x is in $(-\infty, +\infty)$

3. $5x = 2y - 3$

$2y = 5x + 3$

$y = \dfrac{5x + 3}{2}$; x is in $(-\infty, +\infty)$

4. $f^{-1}(x) = \dfrac{5x + 3}{2}$; x is in $(-\infty, +\infty)$ **(Inverse of a function)**

35. $G(x) = |x - 2|$; x is in $(-\infty, +\infty)$, G(x) is in $[0, +\infty)$

1. $y = |x - 2|$; x is in $(-\infty, +\infty)$, $y \geq 0$

2. $x = |y - 2|$; y is in $(-\infty, +\infty)$, $x \geq 0$

3. $y - 2 = x$ or $y - 2 = -x$

$y = x + 2$ or $y = 2 - x$; $x \geq 0$

If x = 1, then y = 3 or 1. Hence, the last equations do not define a function and $f^{-1}(x)$ does not exist. **(Inverse of a function)**

36. A horizontal line is the graph of a function. Every vertical line in the plane containing the horizontal line will intersect the line exactly once. **(Graph of a function)**

37. An oblique line is the graph of a function. Every vertical line in the plane containing the oblique line will intersect the line exactly once. **(Graph of a function)**

38. An ellipse is not the graph of a function. Some vertical lines in the plane containing the ellipse will intersect it twice. **(Graph of a function)**

39. The letter Q is not the graph of a function. Some vertical lines in the plane containing the letter Q will intersect it more than once. **(Graph of a function)**

40. The letter W is the graph of a function. Every vertical line in the plane containing the letter W will interesect it at most once. **(Graph of a function)**

41. **(Graph of a function)**

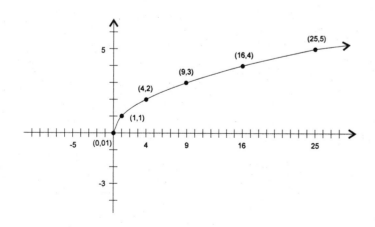

42. (Graph of a function)

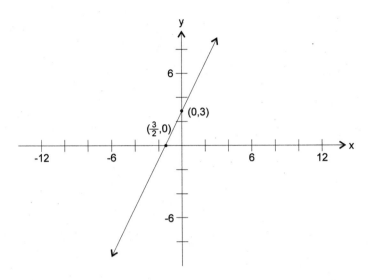

43. (Graph of a function)

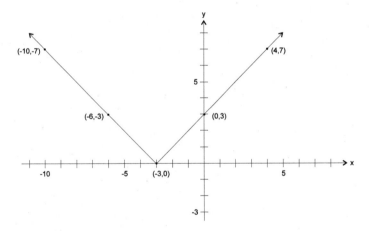

44. **(Graph of a function)**

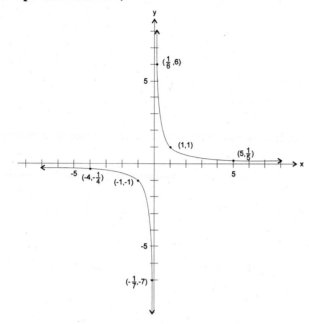

45. **(Graph of a function)**

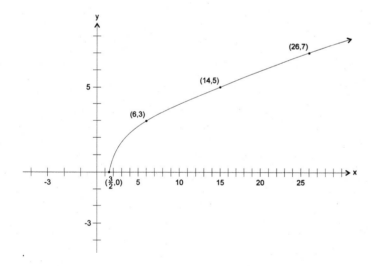

46. **(Graph of a function)**

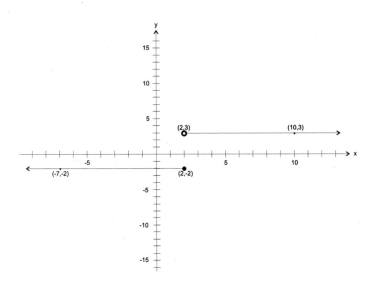

47. **(Graph of a function)**

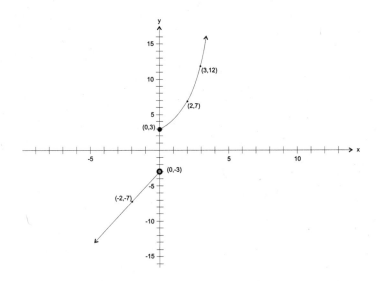

48. f is a linear function. 2x + 7 is of the form ax + b with a = 2 ≠ 0 and b = 7. **(Linear function)**

49. g is a linear function. $\dfrac{5x - 1}{2} = \dfrac{5}{2}x - \dfrac{1}{2}$ is of the form ax + b with a = $\dfrac{5}{2} \neq 0$ and b = $\dfrac{-1}{2}$. **(Linear function)**

50. F is not a linear function. $x^2 + 1$ is not of the form ax + b with a ≠ 0. **(Linear function)**

51. G is a linear function. −5x is of the form ax + b with a = −5 ≠ 0 and b = 0. **(Linear function)**

52. T is a linear function. 1.2x – 3.7 is of the form ax + b with a = 1.2 ≠ 0 and b = –3.7. **(Linear function)**

53. **(Graph of linear function)**

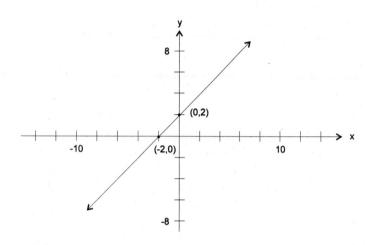

54. **(Graph of linear function)**

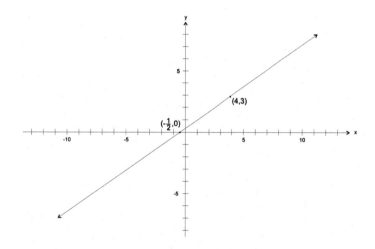

55. **(Graph of linear function)**

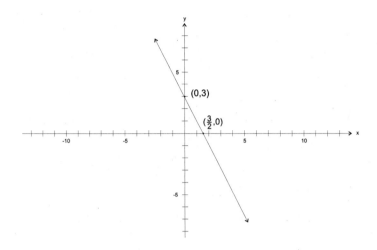

56. For y = f(x) = 2x – 5, slope = 2 and y–intercept = –5. **(Graph of linear function)**

57. For y = h(x) = $\dfrac{3x - 1}{4}$, slope = $\dfrac{3}{4}$ and y–intercept = $\dfrac{-1}{4}$. **(Graph of linear function)**

58. For y = G(x) = –7x, slope = –7 and y–intercept = 0. **(Graph of linear function)**

59. f is a quadratic function. $3x^2 - 5x + 6$ is of the form $ax^2 + bx + c$ with a = 3 ≠ 0, b = –5, and c = 6. **(Quadratic function)**

60. h is a quadratic function. $7 - 9x^2 = -9x^2 + 7$ is of the form $ax^2 + bx + c$ with a = –9 ≠ 0, b = 0, and c = 7. **(Quadratic function)**

61. F is not a quadratic function. 8 – 9x is not of the form $ax^2 + bx + c$ with a ≠ 0. **(Quadratic function)**

62. H is a quadratic function. $2x - 6x^2$ is of the form $ax^2 + bx + c$ with a = –6 ≠ 0, b = 2, and c = 0. **(Quadratic function)**

63. J is not a quadratic function. $x^2 - 5x + 6$ is of the form $ax^2 + bx + c$ with a ≠ 0. However, the expression is under the radical sign. **(Quadratic function)**

64. **(Graph of quadratic function)**

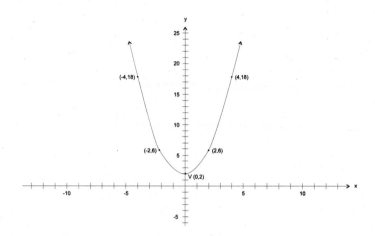

65. **(Graph of quadratic function)**

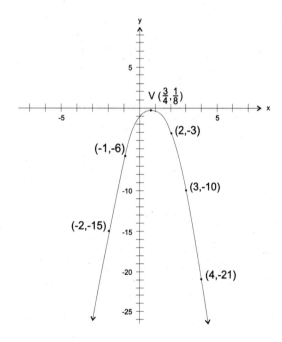

66. **(Graph of quadratic function)**

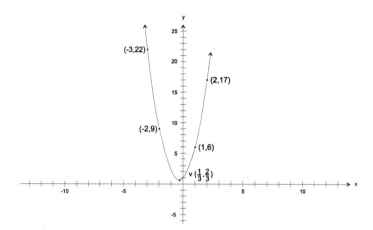

67. **(Graph of quadratic function)**

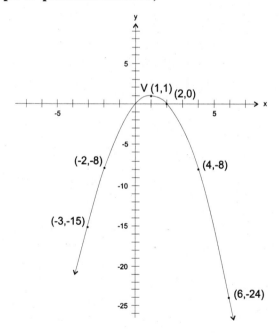

Grade Yourself

Circle the numbers of the questions you missed, then fill in the total incorrect for each topic. If you answered more than three questions incorrectly, you need to focus on that topic. (If a topic has less than three questions and you had at least one wrong, we suggest you study that topic also. Read your textbook, a review book, or ask your teacher for help.)

Subject: Functions

Topic	Question Numbers	Number Incorrect
Domain of a function	1, 2, 3, 4, 5, 6, 7, 8, 9, 10, 11	
Sum of functions	12, 13, 14, 15	
Difference of functions	16, 17, 18, 19	
Product of functions	20, 21, 22, 23	
Quotient of functions	24, 25, 26, 27	
Composition of functions	28, 29, 30, 31	
Inverse of a function	32, 33, 34, 35	
Graph of a function	36, 37, 38, 39, 40, 41, 42, 43, 44, 45, 46, 47	
Linear function	48, 49, 50, 51, 52	
Graph of linear function	53, 54, 55, 56, 57, 58	
Quadratic function	59, 60, 61, 62, 63	
Graph of quadratic function	64, 65, 66, 67	

Limits and Continuity

2

 Test Yourself

2.1 Introduction to Limits

Definition: If f(x) becomes close to a single real number L whenever x is close to but not necessarily equal to c, then we say that the limit of f(x), as x approaches c, is L and write $\lim_{x \to c} f(x) = L$.

(Note that c does not have to be in the domain of f. If c is in the domain of f, and the limit of f(x) as x approaches c exists, then the limit does not have to be equal to f(c).)

Properties of Limits:

    I. If f(x) = k is a constant-valued function, then

$$\lim_{x \to c} f(x) = \lim_{x \to c} (k) = k.$$

    II. If f(x) = x^n for any positive integer n, then

$$\lim_{x \to c} f(x) = \lim_{x \to c} (x^n) = c^n.$$

III. If $\lim_{x \to c} f(x) = L$ and $\lim_{x \to c} g(x) = M$, then

    a. $\lim_{x \to c} (f(x) \pm g(x)) = L \pm M.$

    b. $\lim_{x \to c} (f(x) \cdot g(x)) = L \cdot M$

    c. $\lim_{x \to c} \left(\dfrac{f(x)}{g(x)} \right) = \dfrac{L}{M}$, provided that $M \neq 0$

    d. $\lim_{x \to c} (k \cdot f(x)) = k \cdot \lim_{x \to c} (f(x)) = k \cdot L$

    e. $\lim_{x \to c} \sqrt[n]{f(x)} = \sqrt[n]{\lim_{x \to c} f(x)} = \sqrt[n]{L}$, provided that $\sqrt[n]{L}$ is defined

---

IV. If f is a polynomial function in the variable x, then $\lim_{x \to c} f(x) = f(c)$

In Exercises 1-15, use the properties of limits to determine the limit, if it exists.

1. $\lim_{x \to 2} f(x)$, if $f(x) = x^2$

2. $\lim_{x \to 4} g(x)$, if $g(x) = 13$

3. $\lim_{x \to -1} h(x)$, if $h(x) = x - 3$

4. $\lim_{x \to -2} t(x)$, if $t(x) = -7x^3$

5. $\lim_{x \to 2} F(x)$, if $F(x) = 3x^5 - 2x^2 + x - 1$

6. $\lim_{x \to 0} G(x)$, if $G(x) = \dfrac{x + 1}{x - 2}$

7. $\lim_{x \to -3} H(x)$, if $H(x) = \sqrt{3 - x}$

8. $\lim_{x \to -2} P(x)$, if $P(x) = (2x + 3)(x^2 - 1)$

9. $\lim_{x \to 0} q(x)$, if $q(x) = \dfrac{x - 2}{x^2 + 1}$

10. $\lim_{x \to 0} Q(x)$, if $Q(x) = \dfrac{x}{x - 1}$

11. $\lim\limits_{x \to -1} r(x)$, if $r(x) = 2x^3 - 5x^2 + 2x$

12. $\lim\limits_{x \to 3} S(x)$, if $S(x) = (x^2 - 3x)(x - 3)$

13. $\lim\limits_{x \to 5} w(x)$, if $w(x) = \dfrac{x^2 - 25}{x - 5}$

14. $\lim\limits_{x \to 3} y(x)$, if $y(x) = \dfrac{x^2 - x - 6}{x^2 - 2x - 3}$

15. $\lim\limits_{x \to 1} Y(x)$, if $Y(x) = \dfrac{x^2 - 2x - 1}{x^2 - 5x + 6}$

2.2 One-Sided Limits

Definitions:

1. $\lim\limits_{x \to c^-} f(x) = L$ is called the **left-hand limit** (or the **limit from the left**) if $f(x)$ is close to L whenever x is to the left of c on the real number line and close to c.

2. $\lim\limits_{x \to c^+} f(x) = M$ is called the **right-hand limit** (or the **limit from the right**) if $f(x)$ is close to M whenever x is to the right of c on the real number line and close to c.

Note: $\lim\limits_{x \to c} f(x)$ exists if and only if $\lim\limits_{x \to c^-} f(x)$ exists, $\lim\limits_{x \to c^+} f(x)$ exists, and $\lim\limits_{x \to c^-} f(x) = \lim\limits_{x \to c^+} f(x)$.

---

16. In 1994, first-class postage for mailing a letter was $0.29 for the first ounce or fractional part thereof. For each additional ounce or fractional part thereof, the rate was $.23. Let y = P(x) be the postal function for all $0 \le x \le 5$ where x is the weight, in ounces, and P(x) is the cost.

    a. Graph the function P on [0, 5].

    b. Determine $\lim\limits_{x \to 1.3^-} P(x)$.

    c. Determine $\lim\limits_{x \to 1.3^+} P(x)$.

    d. Determine $\lim\limits_{x \to 1.3} P(x)$.

    e. Compute: P(1.3).

17. Using the function P given in Exercise 16, determine each of the following:

    a. $\lim\limits_{x \to 3^-} P(x)$

    b. $\lim\limits_{x \to 3^+} P(x)$

    c. $\lim\limits_{x \to 3} P(x)$

    d. P(3)

18. Use the graph of the function F (given below) to answer each of the following:

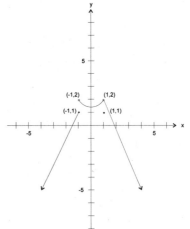

    a. $\lim\limits_{x \to -1^-} F(x)$

    b. $\lim\limits_{x \to -1^+} F(x)$

    c. $\lim\limits_{x \to -1} F(x)$

    d. F(-1)

    e. $\lim\limits_{x \to 1^-} F(x)$

    f. $\lim\limits_{x \to 1^+} F(x)$

    g. $\lim\limits_{x \to 1} F(x)$

    h. F(1)

19. Let $g(x) = \dfrac{x}{|x|}$

    a. What is the domain of g?

    b. Compute: g(2)

    c. Compute: g(–3)

    d. Does $\lim\limits_{x \to 4}$ g(x) exist? Why (not)?

    e. Does $\lim\limits_{x \to 0}$ g(x) exist? Why (not)?

23. Let $h(x) = \dfrac{2x}{x^2 - 1}$

    a. Is h continuous at x = 0? Why (not)?

    b. Is h continuous at x = –1? Why (not)?

24. Let $p(x) = \dfrac{1}{\sqrt{x - 2}}$

    a. Is p continuous at x = 0? Why (not)?

    b. Is p continuous at x = 2? Why (not)?

    c. Is p continuous at x = 6? Why (not)

2.3 Continuity

Definitions:

1. A function f in the variable x is **continuous** at x = a if and only if:

    a. f(a) is defined (i.e., a is in the domain of f),

    b. $\lim\limits_{x \to a}$ f(x) exists, and

    c. $\lim\limits_{x \to a}$ f(x) = f(a).

2. A function is **discontinuous** at a if it is not **continuous** at a.

3. A function is **continuous on an interval** if and only if it is continuous at every point in the interval.

20. Let $f(x) = \begin{cases} 4, & \text{if } x > 3 \\ 3, & \text{if } x = 3 \\ 2, & \text{if } x < 3 \end{cases}$

    a. Is f continuous at x = 0? Why (not)?

    b. Is f continuous at x = 3? Why (not)

21. Let $g(x) = \begin{cases} x^2, & \text{if } x \le 1 \\ x + 1, & \text{if } x > 1 \end{cases}$

    a. Is g continuous at x = –2? Why (not)?

    b. Is g continuous at x = 1? Why (not)?

22. Let $F(x) = \begin{cases} -2, & \text{if } x < -1 \\ x, & \text{if } -1 \le x \le 2 \\ 2, & \text{if } x > 2 \end{cases}$

    a. Is F continuous at x = –1? Why (not)?

    b. Is F continuous at x = 2? Why (not)?

2.4 Properties of Continuity

1. A constant-valued function such as f(x) = 13 is continuous everywhere on its domain (which is the set of all real numbers).

2. A polynomial function is continuous everywhere on its domain (which is the set of all real numbers).

3. A rational function is continuous everywhere on its domain (which is the set of all real numbers for which the denominator of the expression is not equal to 0).

4. If the functions F and G are continuous at x = b, then the following functions are also continuous at x = b:

    a. F + G

    b. F – G

    c. F · G

    d. F/G, provided that G(b) ≠ 0

5. If n is a positive integer, then the function $f(x) = x^n$ is continuous for all real values of *x*.

6. If n is an **odd positive integer** greater than 1, then the function $F(x) = \sqrt[n]{f(x)}$ is continuous wherever *f* is continuous.

7. If n is an **even positive integer**, then the function $G(x) = \sqrt[n]{g(x)}$ is continuous wherever g is continuous and g(x) is nonnegative.

For each of the Exercises 25-31, determine whether the indicated function is continuous at the given values of x.

25. $f(x) = 2x + 3$; $x = -2, 0$

26. $g(x) = \dfrac{x + 1}{x}$; $x = -1, 0$

27. $h(x) = \sqrt{2x - 1}$; $x = 0, 3$

28. $p(x) = 4$; $x = -2, 2$

29. $u(x) = \dfrac{x^2 - 2x + 1}{x^2 - 5x + 6}$; $x = -1, 3$

30. $v(x) = \sqrt[3]{x - 9}$; $x = 3, 9$

31. $y(x) = \dfrac{x^3}{x^3 - 1}$; $x = -1, 1$

In Exercises 32-38, determine all points of discontinuity for each of the given functions.

32. $F(x) = 2x^2 - 5$

33. $H(x) = \dfrac{x + 2}{x^2 - 1}$

34. $J(x) = \sqrt{2x - 7}$

35. $P(x) = \sqrt[5]{1 - 3x}$

36. $V(x) = \dfrac{2x - 11}{3}$

37. $W(x) = -5 + x$

38. $Y(x) = x^2 + 2\sqrt{x}$

Note: Let f be a function in the variable x. If f is continuous at x = c, then $\lim\limits_{x \to c} f(x) = f(c)$.

In Exercises 39-45, evaluate each of the given limits, if the limit exists.

39. $\lim\limits_{x \to 3} (2x + 3)$

40. $\lim\limits_{x \to 0} \dfrac{x^2 - 1}{x + 2}$

41. $\lim\limits_{x \to -2} \sqrt{3x + 7}$

42. $\lim\limits_{x \to 3} \dfrac{4}{x - 5}$

43. $\lim\limits_{x \to 2} (4x - 5)^5$

44. $\lim\limits_{x \to -1} |6x - 7|$

45. $\lim\limits_{x \to 0} \dfrac{2x - 1}{3x^2 + 1}$

 # Check Yourself

1. $\lim\limits_{x \to 2} f(x) = \lim\limits_{x \to 2} (x^2)$

    $= 2^2$ (Property II with n = 2)

    $= 4$ **(Limits)**

2. $\lim\limits_{x \to 4} g(x) = \lim\limits_{x \to 4} (13)$

    $= 13$ (Property I) **(Limits)**

3. $\lim\limits_{x \to -1} h(x) = \lim\limits_{x \to -1} (x - 3)$

$\qquad = \lim\limits_{x \to -1} (x) - \lim\limits_{x \to -1} (3)$ (Property IIIa)

$\qquad = (-1) - \lim\limits_{x \to -1} (3)$ (Property II with n = 1)

$\text{F} \qquad = (-1) - (3)$ (Property I)

$\qquad = -4$

Or, we could use Property IV:

$\lim\limits_{x \to -1} h(x) = \lim\limits_{x \to -1} (x - 3)$

$\qquad = (-1) - (3)$

$\qquad = -4$ **(Limits)**

4. $\lim\limits_{x \to -2} t(x) = \lim\limits_{x \to -2} (-7x^3)$

$\qquad = (-7) \cdot \lim\limits_{x \to -2} (x^3)$ (Property IIId)

$\qquad = (-7)(-2)^3$ (Property II with n = 3)

$\qquad = 56$ **(Limits)**

5. $\lim\limits_{x \to 2} F(x) = \lim\limits_{x \to 2} (3x^5 - 2x^2 + x - 1)$

$\qquad = 3(2)^5 - 2(2)^2 + (2) - 1$ (Property IV)

$\qquad = 96 - 8 + 2 - 1$

$\qquad = 89$ **(Limits)**

6. $\lim\limits_{x \to 0} G(x) = \lim\limits_{x \to 0} \dfrac{x + 1}{x - 2}$

$\qquad = \dfrac{\lim\limits_{x \to 0} (x + 1)}{\lim\limits_{x \to 0} (x - 2)}$ (Property IIIc)

$\qquad = \dfrac{0 + 1}{0 - 2}$ (Property IV)

$\qquad = \dfrac{-1}{2}$ **(Limits)**

7. $\lim\limits_{x \to -3} H(x) = \lim\limits_{x \to -3} \sqrt{3 - x}$

$\qquad = \sqrt{\lim\limits_{x \to -3} (3 - x)}$ (Property IIIe)

$\qquad = \sqrt{3 - (-3)}$ (Property IV)

$\qquad = \sqrt{6}$ **(Limits)**

8. $\lim\limits_{x \to -2} P(x) = \lim\limits_{x \to -2} [(2x + 3)(x^2 - 1)]$

$\qquad = \left[\lim\limits_{x \to -2} (2x + 3)\right]\left[\lim\limits_{x \to -2} (x^2 - 1)\right]$ (Property IIIb)

$\qquad = [2(-2) + 3] [(-2)^2 - 1]$ \qquad (Property IV)

$\qquad = (-1)(3)$

$\qquad = -3$

Or, rewrite $P(x) = (2x + 3)(x^2 - 1)$ as $P(x) = 2x^3 + 3x^2 - 2x - 3$ and use Property IV. (**Limits**)

9. $\lim\limits_{x \to 0} q(x) = \lim\limits_{x \to 0} \dfrac{x - 2}{x^2 + 1}$

$\qquad = \dfrac{\lim\limits_{x \to 0} (x - 2)}{\lim\limits_{x \to 0} (x^2 + 1)}$ \quad (Property IIIc)

$\qquad = \dfrac{(0) - 2}{(0)^2 + 1}$ \qquad (Property IV)

$\qquad = -2$ (**Limits**)

10. $\lim\limits_{x \to 0} Q(x) = \lim\limits_{x \to 0} \dfrac{x}{x - 1}$

$\qquad = \dfrac{\lim\limits_{x \to 0} (x)}{\lim\limits_{x \to 0} (x - 1)}$ \quad (Property IIIc)

$\qquad = \dfrac{0}{0 - 1}$ \qquad (Property IV)

$\qquad = 0$ (**Limits**)

11. $\lim\limits_{x \to -1} r(x) = \lim\limits_{x \to -1} (2x^3 - 5x^2 + 2x)$

$\qquad = 2(-1)^3 - 5(-1)^2 + 2(-1)$ (Property IV)

$\qquad = 2(-1) - 5(1) - 2$

$\qquad = -9$ (**Limits**)

12. $\lim\limits_{x \to 3} S(x) = \lim\limits_{x \to 3} [(x^2 - 3x)(x - 3)]$

$\qquad = [\lim\limits_{x \to 3} (x^2 - 3x)][\lim\limits_{x \to 3} (x - 3)]$

$\qquad = [(3)^2 - 3(3)][3 - 3]$

$\qquad = (0)(0)$

$\qquad = 0$

Or, rewrite $S(x) = (x^2 - 3x)(x - 3)$ as $S(x) = x^3 - 6x^2 + 9x$ and use Property IV. (**Limits**)

13. $\lim\limits_{x \to 5} w(x) = \lim\limits_{x \to 5} \dfrac{x^2 - 25}{x - 5}$

It would appear that Property IIId could be used. However, the limit of the denominator as x→5 is 0. But,

$$\frac{x^2 - 25}{x - 5} = \frac{(x - 5)(x + 5)}{x - 5} = x + 5, \text{ if } x \neq 5$$

Hence, for all x ≠ 5, $\dfrac{x^2 - 25}{x - 5} = x + 5$. We don't care what happens when x = 5, only what happens as x gets

close to 5. Therefore,

$$\lim_{x \to 5} \frac{x^2 - 25}{x - 5} = \lim_{x \to 5} (x + 5) \text{ (Substitution)}$$

$$= 5 + 5 \qquad\qquad \text{(Property IV)}$$

$$= 10 \textbf{ (Limits)}$$

14. $\lim\limits_{x \to 3} y(x) = \lim\limits_{x \to 3} \dfrac{x^2 - x - 6}{x^2 - 2x - 3}$

The limit of the denominator as x→3 is 0. Hence, Property IIId cannot be used. However,

$$\frac{x^2 - x - 6}{x^2 - 2x - 3} = \frac{(x - 3)(x + 2)}{(x - 3)(x + 1)} = \frac{x + 2}{x + 1}, \text{ if } x \neq -1, 3$$

Since we are interested in $\lim\limits_{x \to 3} y(x)$, we don't care what happens when x = 3, only what happens when x

approaches 3. Hence,

$$\lim_{x \to 3} \frac{x^2 - x - 6}{x^2 - 2x - 3} = \lim_{x \to 3} \frac{x + 2}{x + 1} \qquad \text{(Substitution)}$$

$$= \frac{\lim\limits_{x \to 3} (x + 2)}{\lim\limits_{x \to 3} (x + 1)} \qquad\qquad \text{(Property IIId)}$$

$$= \frac{3 + 2}{3 + 1} \qquad\qquad \text{(Property IV)}$$

$$= \frac{5}{4} \qquad \textbf{(Limits)}$$

15. $\lim\limits_{x \to 1} Y(x) = \lim\limits_{x \to 1} \dfrac{x^2 - 2x - 1}{x^2 - 5x + 6}$

$$= \frac{\lim\limits_{x \to 1} (x^2 - 2x - 1)}{\lim\limits_{x \to 1} (x^2 - 5x + 6)} \qquad \text{(Property IIId)}$$

$$= \frac{(1)^2 - 2(1) - 1}{(1)^2 - 5(1) + 6} \qquad \text{(Property IV)}$$

$$= -1 \textbf{ (Limits)}$$

16. a.

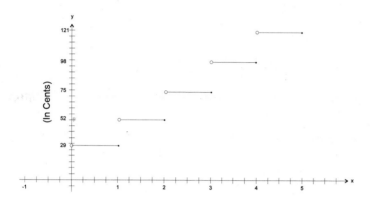

 b. $\lim\limits_{x \to 1.3^-} P(x) = \$.52$

 c. $\lim\limits_{x \to 1.3^+} P(x) = \$.52$

 d. $\lim\limits_{x \to 1.3} P(x) = \$.52$

 e. $P(1.3) = \$.52$ **(One–sided limits)**

17. a. $\lim\limits_{x \to 3^-} P(x) = \$.75$

 b. $\lim\limits_{x \to 3^+} P(x) = \$.98$

 c. $\lim\limits_{x \to 3} P(x)$ does not exist since $\lim\limits_{x \to 3^-} P(x) \neq \lim\limits_{x \to 3^+} P(x)$.

 d. $P(3) = \$.75$ **(One–sided limits)**

18. a. $\lim\limits_{x \to -1^-} F(x) = 1$

 b. $\lim\limits_{x \to -1^+} F(x) = 2$

 c. $\lim\limits_{x \to -1} F(x)$ does not exist since $\lim\limits_{x \to -1^-} F(x) \neq \lim\limits_{x \to -1^+} F(x)$.

 d. $F(-1) = 1$

 e. $\lim\limits_{x \to 1^-} F(x) = 2$

 f. $\lim\limits_{x \to 1^+} F(x) = 2$

 g. $\lim\limits_{x \to 1} F(x) = 2$

 h. $F(1) = 1$ **(One–sided limits)**

19. a. The domain of g is the set of all nonzero real numbers.

b. $g(2) = \dfrac{2}{|2|} = \dfrac{2}{2} = 1$

c. $g(-3) = \dfrac{-3}{|-3|} = \dfrac{-3}{3} = -1$

d. $\lim\limits_{x\to 4^-} g(x) = 1$ and $\lim\limits_{x\to 4^+} g(x) = 1$. Therefore, $\lim\limits_{x\to 4} g(x)$ exists (and is equal to 1).

e. $\lim\limits_{x\to 0^-} g(x) = -1$ and $\lim\limits_{x\to 0^+} g(x) = 1$. Therefore, $\lim\limits_{x\to 0} g(x)$ does not exist (since $-1 \neq 1$). **(One–sided limits)**

20. $f(x) = \begin{cases} 4, \text{ if } x > 3 \\ 3, \text{ if } x = 3 \\ 2, \text{ if } x < 3 \end{cases}$

a. i) Is f(0) defined? Yes, f(0) = 2.

ii) Does $\lim\limits_{x\to 0} f(x)$ exist?

Yes, $\lim\limits_{x\to 0} f(x) = \lim\limits_{x\to 0}(2) = 2$.

iii) Does $\lim\limits_{x\to 0} f(x) = f(0)$?

Yes, $\lim\limits_{x\to 0} f(x) = 2 = f(0)$.

Therefore, f is continuous at x = 0.

b. i) Is f(3) defined? Yes, f(3) = 3.

ii) Does $\lim\limits_{x\to 3} f(x)$ exist?

$\lim\limits_{x\to 3^-} f(x) = \lim\limits_{x\to 3^-}(2) = 2$

$\lim\limits_{x\to 3^-} f(x) = \lim\limits_{x\to 3^+}(4) = 4$

Since $\lim\limits_{x\to 3^-} f(x) = 2 \neq 4 = \lim\limits_{x\to 3^+} f(x)$, $\lim\limits_{x\to 3} f(x)$ does not exist.

Therefore, f is not continuous at x = 3. **(Continuity)**

21. $g(x) = \begin{cases} x^2, \text{ if } x \leq 1 \\ x+1, \text{ if } x > 1 \end{cases}$

a. i) Is g(–2) defined? Yes, $g(-2) = (-2)^2 = 4$.

ii) Does $\lim\limits_{x\to -2} g(x)$ exist?

Yes, $\lim\limits_{x\to -2} g(x) = \lim\limits_{x\to -2}(x^2) = 4$.

iii) Does $\lim\limits_{x\to -2} g(x) = g(-2)$?

Yes, $\lim\limits_{x\to -2} g(x) = 4 = g(-2)$.

Therefore, g is continuous at x = –2.

b.   i) Is g(1) defined? Yes, $g(1) = (1)^2 = 1$.

   ii) Does $\lim\limits_{x \to 1} g(x)$ exist?

   No, $\lim\limits_{x \to 1^-} g(x) = \lim\limits_{x \to 1^-} (x^2) = (1^2) = 1$,

   $\lim\limits_{x \to 1^+} g(x) = \lim\limits_{x \to 1^+} (x + 1) = 1 + 1 = 2$, and $1 \neq 2$.

Therefore, g is not continuous at x = 1. **(Continuity)**

22. $F(x) = \begin{cases} -2, \text{if } x < -1 \\ x, \text{ if } -1 \le x \le 2 \\ 2, \text{ if } x > 2 \end{cases}$

a.   i) Is F(–1) defined? Yes, F(–1) = –1.

   ii) Does $\lim\limits_{x \to -1} F(x)$ exist?

   $\lim\limits_{x \to -1^-} F(x) = \lim\limits_{x \to -1^-} (-2) = -2$

   $\lim\limits_{x \to -1^+} F(x) = \lim\limits_{x \to -1^+} (x) = -1$

   Since $\lim\limits_{x \to -1^-} F(x) = -2 \neq -1 = \lim\limits_{x \to -1^+} F(x)$, $\lim\limits_{x \to -1} F(x)$ does not exist.

Therefore, F is not continuous at x = –1.

b.   i) Is F(2) defined? Yes, F(2) = 2.

   ii) Does $\lim\limits_{x \to 2} F(x)$ exist?

   $\lim\limits_{x \to 2^-} F(x) = \lim\limits_{x \to 2^-} (x) = 2$

   $\lim\limits_{x \to 2^+} F(x) = \lim\limits_{x \to 2^+} (2) = 2$

Since $\lim\limits_{x \to 2^-} F(x) = 2 = \lim\limits_{x \to 2^+} F(x)$, $\lim\limits_{x \to 2} F(x)$ does exist.

   iii) Does $\lim\limits_{x \to 2} F(x) = F(2)$?

   Yes, $\lim\limits_{x \to 2} F(x) = 2 = F(2)$.

Therefore, F is continuous at x = 2. **(Continuity)**

23. $h(x) = \dfrac{2x}{x^2 - 1}$

a.   i) Is h(0) defined? Yes, $h(0) = \dfrac{2(0)}{(0)^2 - 1} = \dfrac{0}{-1} = 0$.

   ii) Does $\lim\limits_{x \to 0} h(x)$ exist?

$$\text{Yes, } \lim_{x \to 0} h(x) = \lim_{x \to 0} \frac{2x}{x^2 - 1} = \frac{2(0)}{(0) - 1} = 0.$$

iii) Does $\lim_{x \to 0} h(x) = h(0)$?

$$\text{Yes, } \lim_{x \to 0} h(x) = 0 = h(0).$$

Therefore, h is continuous at x = 0.

b. i) Is h(−1) defined? No, $h(-1) = \frac{2(-1)}{(-1)^2 - 1} = \frac{-2}{0}$ which is not defined.

Therefore, h is not continuous at x = −1. **(Continuity)**

24. $p(x) = \frac{1}{\sqrt{x - 2}}$

a. i) Is p(0) defined? No, $p(0) = \frac{1}{\sqrt{0 - 2}} = \frac{1}{\sqrt{-2}}$ which is not a real number.

Therefore, p is not continuous at x = 0.

b. i) Is p(2) defined? No, $p(2) = \frac{1}{\sqrt{2 - 2}} = \frac{1}{\sqrt{0}} = \frac{1}{0}$ which is not defined.

Therefore, p is not continuous at x = 2.

c. i) Is p(6) defined? Yes, $p(6) = \frac{1}{\sqrt{6 - 2}} = \frac{1}{\sqrt{4}} = \frac{1}{2}.$

ii) Does $\lim_{x \to 6} p(x)$ exist?

$$\text{Yes, } \lim_{x \to 6} p(x) = \lim_{x \to 6} \frac{1}{\sqrt{x - 2}} = \frac{1}{\sqrt{\lim_{x \to 6}(x - 2)}} = \frac{1}{\sqrt{6 - 2}} = \frac{1}{\sqrt{4}} = \frac{1}{2}$$

iii) Does $\lim_{x \to 6} p(x) = p(6)$?

$$\text{Yes, } \lim_{x \to 6} p(x) = \frac{1}{2} = p(6).$$

Therefore, p is continuous at x = 6. **(Continuity)**

25. f(x) = 2x + 3 is a polynomial function which is continuous everywhere on its domain (which is the set of all real numbers). Hence, f is continuous at −2 and 0. **(Continuity)**

26. $g(x) = \frac{x + 1}{x}$ is a rational function which is continuous everywhere on its domain. The domain of g is $(-\infty, 0) \cup (0, +\infty)$. Hence, g is continuous at −1 but is discontinuous at 0. **(Continuity)**

27. $h(x) = \sqrt{2x - 1}$ is continuous whenever 2x − 1 is continuous and nonnegative, which is for all x in $[\frac{1}{2}, +\infty)$. Hence, h is discontinuous at 0 but is continuous at 3. **(Continuity)**

28. p(x) = 4 is a constant–valued function which is continuous everywhere on the set of all real numbers. Hence, p is continuous at −2 and 2. **(Continuity)**

29. $u(x) = \dfrac{x^2 - 2x + 1}{x^2 - 5x + 6} = \dfrac{x^2 - 2x + 1}{(x - 2)(x - 3)}$ is a rational function which is continuous everywhere on its domain.
The domain of u is $(-\infty, 2) \cup (2, 3) \cup (3, +\infty)$. Hence, u is continuous at −1 but is discontinuous at 3. (**Continuity**)

30. $v(x) = \sqrt[3]{x - 9}$ is continuous wherever x − 9 is continuous. But, x − 9 is continuous everywhere on the set of all real numbers. Hence, v is continuous at 3 and 9. (**Continuity**)

31. $y(x) = \dfrac{x^3}{x^3 - 1}$ is a rational function which is continuous everywhere on its domain. The domain of y is $(-\infty, 1) \cup (1, +\infty)$, Hence, y is continuous at −1 but is discontinuous at 1. (**Continuity**)

32. $F(x) = 2x^2 - 5$ is a polynomial function which is continuous on the set of all real numbers. Hence, there are no points of discontinuity. (**Continuity**)

33. $H(x) = \dfrac{x + 2}{x^2 - 1} = \dfrac{x + 2}{(x - 1)(x + 1)}$ is a rational function which is continuous everywhere on its domain. The domain of H is $(-\infty, -1) \cup (-1, 1) \cup (1, +\infty)$. Hence, H is discontinuous at −1 and 1. (**Continuity**)

34. $J(x) = \sqrt{2x - 7}$ is continuous wherever 2x − 7 is continuous and nonnegative, which is the set $[\frac{7}{2}, +\infty)$.
Hence, J is discontinuouus on $(-\infty, \frac{7}{2})$. (**Continuity**)

35. $P(x) = \sqrt[5]{1 - 3x}$ is continuous wherever 1 − 3x is continuous. But, 1 − 3x is continuous everywhere on the set of all real numbers. Hence, there are no points of discontinuity for P. (**Continuity**)

36. $V(x) = \dfrac{2x - 11}{3} = \dfrac{2}{3}x - \dfrac{11}{3}$ is a polynomial function which is continuous everywhere on the set of all real numbers. Hence, V has no points of discontinuity. (**Continuity**)

37. $W(x) = -5 + x$ is a polynomial function which is continuous everywhere on the set of all real numbers. Hence, V has no points of discontinuity. (**Continuity**)

38. $Y(x) = x^2 + 2\sqrt{x}$ is continuous wherever x^2 is continuous *and* $2\sqrt{x}$ is continuous *and* $x \geq 0$. x^2 is continuous everywhere on the set of all real numbers. $2\sqrt{x}$ is continuous on $[0, +\infty)$. Therefore, Y is continuous on $[0, +\infty)$. Hence, Y is undefined and therefore discontinuous on $(-\infty, 0)$. (**Continuity**)

39. Let $f(x) = 2x + 3$, which is continuous at x = 3. Therefore, $\lim_{x \to 3} (2x + 3) = f(3) = 2(3) + 3 = 9$. (**Limits**)

40. Let $g(x) = \dfrac{x^2 - 1}{x + 2}$, which is continuous at x = 0. Therefore, $\lim_{x \to 0} \dfrac{x^2 - 1}{x + 2} = g(0) = \dfrac{(0)^2 - 1}{0 + 2} = \dfrac{-1}{2}$.
(**Limits**)

41. Let $j(x) = \sqrt{3x + 7}$, which is continuous at x = −2. Therefore, $\lim_{x \to -2} \sqrt{3x + 7} = j(-2) = \sqrt{3(-2) + 7} = \sqrt{-6 + 7} = 1$. (**Limits**)

42. Let $p(x) = \dfrac{4}{x-5}$, which is continuous at x = 3. Therefore, $\lim\limits_{x \to 3} \dfrac{4}{x-5} = p(3) = \dfrac{4}{3-5} = -2$. **(Limits)**

43. Let $H(x) = (4x - 5)^5$, which is continuous at x = 2. Therefore,

    $\lim\limits_{x \to 2} (4x - 5)^5 = H(2) = [4(2) - 5]^5 = (3)^5 = 243$. **(Limits)**

44. Let $J(x) = |\,6x - 7\,|$, which is continuous at x = −1. Therefore,

    $\lim\limits_{x \to -1} |\,6x - 7\,| = J(-1) = |\,6(-1) - 7\,| = 13$. **(Limits)**

45. Let $T(x) = \dfrac{2x - 1}{3x^2 + 1}$, which is continuous at x = 0. Therefore,

    $\lim\limits_{x \to 0} \dfrac{2x - 1}{3x^2 + 1} = T(0) = \dfrac{2(0) - 1}{3(0)^2 + 1} = \dfrac{-1}{1} = -1$. **(Limits)**

Grade Yourself

Circle the numbers of the questions you missed, then fill in the total incorrect for each topic. If you answered more than three questions incorrectly, you need to focus on that topic. (If a topic has less than three questions and you had at least one wrong, we suggest you study that topic also. Read your textbook, a review book, or ask your teacher for help.)

Subject: Limits and Continuity

Topic	Question Numbers	Number Incorrect
Limits	1, 2, 3, 4, 5, 6, 7, 8, 9, 10, 11, 12, 13, 14, 15, 39, 40, 41, 42, 43, 44, 45	
One-sided limits	16, 17, 18, 19	
Continuity	20, 21, 22, 23, 24, 25, 26, 27, 28, 29, 30, 31, 32, 33, 34, 35, 36, 37, 38	

The Derivative

3

 Test Yourself

3.1 Introduction

Definitions: Let $y = f(x)$ be a continuous function. Let $P_1(x_1, f(x_1))$ and $P_2(x_2, f(x_2))$ be two points on the graph of f such that $x_2 = x_1 + h$. Then, the expression

$$\frac{f(x_1 + h) - f(x_1)}{h}$$

is called a **difference quotient** and represents the **slope of the secant line** between the two points P_1 and P_2 on the graph of f.

Definition: Let $y = f(x)$ be a continuous function. Then, the **average rate of change** for $f(x)$ from $x = x_1$ to $x = x_2 = x_1 + h$ is given by

Average Rate of Change $= \dfrac{f(x_1 + h) - f(x_1)}{h}$ $(h \neq 0)$.

In Exercises 1-6, compute the indicated difference quotient for each of the given functions.

1. $f(x) = 2x + 5$; $\dfrac{f(1 + h) - f(1)}{h}$

2. $f(x) = x^2$; $\dfrac{f(1 + h) - f(1)}{h}$

3. $f(x) = x^2 + 2x$; $\dfrac{f(2 + h) - f(2)}{h}$

4. $f(x) = 4x - 5$; $\dfrac{f(-3 + h) - f(3)}{h}$, if $h = 1$

5. $f(x) = 2x^2 - 1$; $\dfrac{f(1 + h) - f(1)}{h}$, $h = 0.5$

6. $f(x) = x^3 - 1$; $\dfrac{f(0 + h) - f(0)}{h}$, if $h = 0.02$

In Exercises 7-8, let $y = f(x) = x^2 + 1$.

7. Determine the slope of the secant line between the two points $(0, f(0))$ and $(3, f(3))$.

8. Determine the slope of the secant line between the two points $(-1.5, f(-1.5))$ and $(2, f(2))$.

In Exercises 9–10, let $y = g(x) = 1 - x^3$.

9. Determine the slope of the secant line between the two points $(-2, g(-2))$ and $(1, g(1))$.

10. Determine the slope of the secant line between the two points $(3, g(3))$ and $(5, g(5))$.

In Exercises 11-12, let $y = G(x) = \dfrac{2}{x - 1}$.

11. Determine the slope of the secant line between the two points $(-5, G(-5))$ and $(0, G(0))$.

12. Determine the slope of the secant line between the two points $(2.1, G(2.1))$ and $(4.2, G(4.2))$.

13. Determine the average rate of change of $f(x) = x^2 + x$, with respect to x, from $x = -2$ to $x = 0$.

14. Determine the average rate of change of $h(x) = \sqrt{1 - 2x}$, with respect to x, from $x = -4$ to $x = 0$.

15. Determine the average rate of change of $t(x) = 3x^2 - 2x + 1$, with respect to x, from $x = 1.1$ to $x = 2.2$.

Definition: Let $y = f(x)$ be a continuous function. The **derivative of f(x) with respect to x**, denoted by $f'(x)$ is defined by

$$f'(x) = \lim_{h \to 0} \frac{f(x + h) - f(x)}{h}$$

provided that the limit exists. Geometrically, the derivative, $f'(x)$, gives the slope of the tangent line to the graph of f at the point $(x, f(x))$.

In Exercises 16-22, use the definition for the derivative to determine a simplified expression for $f'(x)$ for each of the functions given.

16. $f(x) = 2x$

17. $f(x) = -7$

18. $f(x) = \sqrt{x - 2}$

19. $f(x) = \dfrac{3}{x - 2}$

20. $f(x) = x^3 - 1$

21. $f(x) = x^2 + x + 1$

22. $f(x) = 2x - x^3$

23. Determine the equation of the line tangent to the graph of $y = f(x) = 2x^2 + 1$ at the point where $x = 1$.

24. Determine the equation of the line tangent to the graph of $y = f(x) = x^3 - 2$ at the point where $x = 0$.

25. Determine the equation of the line tangent to the graph of $y = f(x) = 8$ at the point where $x = 3$.

3.2 Some Rules for Differentiation

(i) If $y = c$ is a constant–valued function, then $\dfrac{dy}{dx} = \dfrac{d(c)}{dx} = 0$ for all real values of x. (That is, the derivative of a constant with respect to a variable is always 0.)

(ii) If $y = x$, then $\dfrac{dy}{dx} = \dfrac{d(x)}{dx} = 1$ for all real values of x. (That is, the derivative of a variable with respect to the variable is always 1.)

(iii) If $y = x^n$, where n is any real number, then $\dfrac{dy}{dx} = \dfrac{d(x^n)}{dx} = nx^{n-1}$ for all real values of x for which x^n is defined.

(iv) If $y = c \cdot f(x)$, then $\dfrac{dy}{dx} = \dfrac{d(c \cdot f(x))}{dx} = c \cdot f'(x)$ where c is a *constant factor.*

(v) The derivative of the sum (or difference) of two functions is the sum (or difference) of their derivatives, provided all the derivatives involved exist. That is,

$$\frac{d}{dx}[f(x) \pm g(x)] = f'(x) \pm g'(x).$$

26. Use the definition of a derivative to prove Rule (i).

27. Use the definition of a derivative to prove Rule (ii).

28. Use the definition of a derivative to prove Rule (iii) with $n = 2$.

In Exercises 29-34, use the rules of this section to determine a simplified expression for $f'(x)$ for each of the functions given.

29. $f(x) = 13$

30. $f(x) = 2x + 7$

31. $f(x) = x^7$

32. $f(x) = x^{-5}$

33. $f(x) = -7x$

34. $f(x) = x^2 + 2x + \sqrt{x}$

3.3 Derivatives of Products and Quotients

Product Rule: If $y = f(x) \cdot g(x)$, where f and g are differentiable functions in x, then

$$\frac{dy}{dx} = \frac{d}{dx}[f(x) \cdot g(x)] = f(x) \cdot g'(x) + g(x) \cdot f'(x)$$

on the common domain of the domains of f' and g'.

That is, the derivative of the product of two functions is the first function times the derivative of the second plus the second function times the derivative of the first.

In Exercises 35-41, use the product rule to determine a simplified expression for $f'(x)$ for each of the given functions.

35. $f(x) = (1 + 2x)(3 - x)$

36. $f(x) = (x^2 - 4)(3x + 8)$

37. $f(x) = \sqrt[3]{x}\,(2 - x^4)$

38. $f(x) = (4x^3 - 1)(x^2 + 7)$

39. $f(x) = -5x^7(4x^8 - 5)$

40. $f(x) = (2x^2 - 1)(3 - 5x^3)$

41. $f(x) = (9 - \sqrt[5]{x}\,)(x^6 + 3)$

Quotient Rule: If $y = \dfrac{f(x)}{g(x)}$ where f and g are differentiable functions in x, then

$$\frac{dy}{dx} = \frac{d}{dx}\left(\frac{f(x)}{g(x)}\right) = \frac{g(x) \cdot f'(x) - f(x) \cdot g'(x)}{[g(x)]^2}$$

provided that $g(x) \neq 0$.

That is, the derivative of the quotient of two functions is equal to the denominator times the derivative of the numerator minus the numerator times the derivative of the denominator, all divided by the square of the denominator, provided that the denominator is not equal to 0.

In Exercises 42-48, use the quotient rule to determine a simplified expression for $f'^1(x)$ for each of the given functions.

42. $f(x) = \dfrac{x + 1}{x - 2}$

43. $f(x) = \dfrac{\sqrt{x}}{x + 3}$

44. $f(x) = \dfrac{4 - (1/x)}{3x^2}$

45. $f(x) = \dfrac{2x - 3}{5 - 4x}$

46. $f(x) = \dfrac{x - 3}{x^2 - 3x - 4}$

47. $f(x) = \dfrac{x^3 - 1}{x^4 + 1}$

48. $f(x) = \dfrac{3x - 1}{2 - \sqrt{x}}$

In Exercises 49-51, use appropriate rules to determine a simplified expression for $f'(x)$ for each of the following.

49. $f(x) = \dfrac{(2x - 3)(x + 1)}{x^2 + 1}$

50. $f(x) = \dfrac{-3(2 - 5x^6)}{8}$

51. $f(x) = \dfrac{2x^3(3x - 5)}{1 - \sqrt{x}}$

3.4 The Chain Rule and General Power Rule

Chain Rule: If $y = f(u)$ is a differentiable function of u, and $u = g(x)$ is a differentiable function of x, then $y = f(g(x))$ is a differentiable function of x and

$$\frac{dy}{dx} = \frac{d}{dx}(f(g(x))) = f'(g(x)) \cdot g'(x) = f'(u) \cdot g'(x)$$

or, in equivalent form, $\dfrac{dy}{dx} = \dfrac{dy}{du} \cdot \dfrac{du}{dx}$

General Power Rule (A Special Case of the Chain Rule): If $y = [u(x)]^n$, where u is a differentiable function in x and n is a real number, then

$$\frac{dy}{dx} = \frac{d}{dx}[u(x)]^n = n[u(x)]^{n-1}\frac{d(u(x))}{dx}$$

or, in equivalent form,

$$\frac{dy}{dx} = nu^{n-1}\frac{du}{dx} \text{ (where } u = u(x)\text{).}$$

In Exercises 52-58, determine a simplified expression for f '(x) for each of the given functions.

52. $f(x) = (2x - 3)^4$

53. $f(x) = \sqrt{3 - 2x}$

54. $f(x) = (x^3 + 1)^7$

55. $f(x) = (7x - 1)^{-3}$

56. $f(x) = (8x^6 + 4x - 1)^2$

57. $f(x) = \sqrt[3]{3x - 5}$

58. $f(x) = (3x^5 - 4x^3 + x - 5)^3$

In Exercises 59-65, use any combination of the differentiation rules to determine a simplified expression for f '(x) for each of the given functions.

59. $f(x) = (x^2 + 1)^3(2x - 1)$

60. $f(x) = \frac{2x(x - 3)}{x + 1}$

61. $f(x) = \frac{-3x(x + 5)^3}{5\sqrt{x + 2}}$

62. $f(x) = x^3\sqrt{3x^2 + 5}$

63. $f(x) = \frac{\sqrt{x} + 3}{(x^2 + 2)^3}$

64. $f(x) = \frac{-5x^3}{\sqrt{x^2 + 4x + 4}}$

65. $f(x) = \sqrt[3]{(1 - x)^4}\ (x^2 - 3)^2$

 # Check Yourself

1. $f(x) = 2x + 5$

$$\frac{f(1 + h) - f(1)}{h} = \frac{[2(1 + h) + 5] - [2(1) + 5]}{h} = \frac{2 + 2h + 5 - 7}{h}$$

$$= \frac{2h}{h} = 2 \text{ (if } h \neq 0) \quad \textbf{(Difference quotient)}$$

2. $f(x) = x^2$

$$\frac{f(1 + h) - f(1)}{h} = \frac{(1 + h)^2 - (1)^2}{h} = \frac{(1 + 2h + h^2) - 1}{h} = \frac{2h + h^2}{h}$$

$$= 2 + h \text{ (if } h \neq 0) \quad \textbf{(Difference quotient)}$$

3. $f(x) = x^2 + 2x$

$$\frac{f(2 + h) - f(2)}{h} = \frac{[(2 + h)^2 + 2(2 + h)] - [(2)^2 + 2(2)]}{h}$$

$$= \frac{(4 + 4h + h^2 + 4 + 2h) - (4 + 4)}{h} = \frac{6h + h^2}{h}$$

$$= 6 + h \text{ (if } h \neq 0) \quad \textbf{(Difference quotient)}$$

4. $f(x) = 4x - 5; h = 1$

$$\frac{f(-3 + h) - f(-3)}{h} = \frac{f(-3 + 1) - f(-3)}{1} = \frac{f(-2) - f(-3)}{1}$$

$$= \frac{[4(-2) - 5] - [4(-3) - 5]}{1} = \frac{(-8 - 5) - (-12 - 5)}{1}$$

$$= \frac{(-13) - (-17)}{1} = 4 \quad \textbf{(Difference quotient)}$$

5. $f(x) = 2x^2 - 1; h = 0.5$

$$\frac{f(1 + h) - f(1)}{h} = \frac{f(1 + 0.5) - f(1)}{0.5} = \frac{f(1.5) - f(1)}{0.5}$$

$$= \frac{[2(1.5)^2 - 1] - [2(1)^2 - 1]}{0.5} = \frac{[2(2.25) - 1] - [2(1) - 1]}{0.5} = \frac{(4.5 - 1) - (2 - 1)}{0.5} = \frac{3.5 - 1}{0.5}$$

$$= \frac{2.5}{0.5} = 5 \quad \textbf{(Difference quotient)}$$

6. $f(x) = x^3 - 1; h = 0.02$

$$\frac{f(0 + h) - f(0)}{h} = \frac{f(0 + 0.02) - f(0)}{0.02} = \frac{f(0.02) - f(0)}{0.02}$$

$$= \frac{[(0.02)^3 - 1] - [(0)^3 - 1]}{0.02} = \frac{(0.000008 - 1) - (-1)}{0.02}$$

$$= \frac{0.000008}{0.02} = 0.0004 \quad \textbf{(Difference quotient)}$$

7. $y = f(x) = x^2 + 1$

$$m_{sec} = \frac{f(3) - f(0)}{3 - 0} = \frac{[(3)^2 + 1] - [(0)^2 + 1]}{3} = \frac{10 - 1}{3} = \frac{9}{3} = 3 \quad \textbf{(Slope of secant line)}$$

8. $y = f(x) = x^2 + 1$

$$m_{sec} = \frac{f(2) - f(-1.5)}{2 - (-1.5)} = \frac{[(2)^2 + 1] - [(1.5)^2 + 1]}{3.5}$$

$$= \frac{5 - 3.25}{3.5} = \frac{1.75}{3.5} = 0.5 \quad \textbf{(Slope of secant line)}$$

9. $y = g(x) = 1 - x^3$

$$m_{sec} = \frac{g(1) - g(-2)}{1 - (-2)} = \frac{[1 - (1)^3] - [1 - (-2)^3]}{3} = \frac{0 - 9}{3} = -3 \quad \textbf{(Slope of secant line)}$$

10. $y = g(x) = 1 - x^3$

$$m_{sec} = \frac{g(5) - g(3)}{5 - 3} = \frac{[1 - (5)^3] - [1 - (3)^3]}{2}$$

$$= \frac{(1 - 125) - (1 - 27)}{2} = \frac{-124 + 26}{2} = \frac{-98}{2} = -49 \quad \textbf{(Slope of secant line)}$$

11. $y = G(x) = \dfrac{2}{x-1}$

$$m_{sec} = \frac{G(0) - G(-5)}{0 - (-5)} = \frac{\left(\dfrac{2}{0-1}\right) - \left(\dfrac{2}{-5-1}\right)}{5} = \frac{-2 - \left(\dfrac{-1}{3}\right)}{5} = \frac{-1}{3} \quad \textbf{(Slope of secant line)}$$

12. $y = G(x) = \dfrac{2}{x-1}$

$$m_{sec} = \frac{G(4.2) - G(2.1)}{4.2 - 2.1} = \frac{\left(\dfrac{2}{4.2-1}\right) - \left(\dfrac{2}{2.1-1}\right)}{2.1} = \frac{\dfrac{2}{3.2} - \dfrac{2}{1.1}}{2.1}$$

$$\approx \frac{0.625 - 1.82}{2.1} \approx -0.57 \quad \textbf{(Slope of secant line)}$$

13. $f(x) = x^2 + x$

$$\text{Avg R. of C.} = \frac{f(0) - f(-2)}{0 - (-2)} = \frac{[(0)^2 + 0] - [(-2)^2 + (-2)]}{2}$$

$$= \frac{0 - 2}{2} = -1 \quad \textbf{(Average rate of change)}$$

14. $h(x) = \sqrt{1 - 2x}$

$$\text{Avg R. of C.} = \frac{h(0) - h(-4)}{0 - (-4)} = \frac{\sqrt{1 - 2(0)} - \sqrt{1 - 2(-4)}}{4}$$

$$= \frac{\sqrt{1} - \sqrt{9}}{4} = \frac{1 - 3}{4} = -0.5 \quad \textbf{(Average rate of change)}$$

15. $t(x) = 3x^2 - 2x + 1$

$$\text{Avg R. of C.} = \frac{t(2.2) - t(1.1)}{2.2 - 1.1}$$

$$= \frac{[3(2.2)^2 - 2(2.2) + 1] - [3(1.1)^2 - 2(1.1) + 1]}{1.1}$$

$$= \frac{[3(4.84) - 4.4 + 1] - [3(1.21) - 2.2 + 1]}{1.1}$$

$$= \frac{11.12 - 2.43}{1.1} = \frac{8.69}{1.1} = 7.9 \quad \textbf{(Average rate of change)}$$

16. $f(x) = 2x$ for all real values of x

$$f'(x) = \lim_{h \to 0} \frac{f(x+h) - f(x)}{h}$$

$$= \lim_{h \to 0} \frac{2(x+h) - 2x}{h}$$

$$= \lim_{h \to 0} \frac{2x + 2h - 2x}{h}$$

$$= \lim_{h \to 0} \frac{2h}{h}$$

$$= \lim_{h \to 0} (2) \qquad \text{(since } h \neq 0)$$

$$= 2 \text{ for all real values of x.}$$

Therefore, if $f(x) = 2x$, then $f'(x) = 2$ for all real values of x. **(Derivative)**

17. $f(x) = -7$ for all real values of x

$$f'(x) = \lim_{h \to 0} \frac{f(x + h) - f(x)}{h}$$

$$= \lim_{h \to 0} \frac{(-7) - (-7)}{h}$$

$$= \lim_{h \to 0} \left(\frac{0}{h} \right)$$

$$= \lim_{h \to 0} (0) \qquad \text{since } (h \neq 0)$$

$$= 0 \text{ for all real values of x.}$$

Therefore, if $f(x) = -7$, then $f'(x) = 0$ for all real values of x.   **(Derivative)**

18. $f(x) = \sqrt{x - 2}$ for all real valules of $x \geq 2$

$$f'(x) = \lim_{h \to 0} \frac{f(x + h) - f(x)}{h}$$

$$= \lim_{h \to 0} \frac{\sqrt{(x + h) - 2} - \sqrt{x - 2}}{h}$$

But, $\dfrac{\sqrt{(x + h) - 2} - \sqrt{x - 2}}{h}$

$$= \frac{\sqrt{(x + h) - 2} - \sqrt{x - 2}}{h} \cdot \frac{\sqrt{(x + h) - 2} + \sqrt{x - 2}}{\sqrt{(x + h) - 2} + \sqrt{x - 2}}$$

$$= \frac{[(x + h - 2)] - (x - 2)}{h[\sqrt{(x + h)} - 2 + \sqrt{x - 2}} = \frac{h}{h[\sqrt{(x + h) - 2} + \sqrt{x - 2}]}$$

$$= \frac{1}{\sqrt{(x + h) - 2} + \sqrt{x - 2}} \qquad \text{since } (h \neq 0)$$

Hence, $= \lim_{h \to 0} \dfrac{\sqrt{(x + h) - 2} - \sqrt{x - 2}}{h}$

$$= \lim_{h \to 0} \frac{1}{\sqrt{(x + h) - 2} + \sqrt{x - 2}}$$

$$= \frac{1}{\sqrt{x - 2} + \sqrt{x - 2}} = \frac{1}{2\sqrt{x - 2}} \text{ for all } x > 2$$

Therefore, if $f(x) = \sqrt{x - 2}$ then $f'(x) = \dfrac{1}{2\sqrt{x - 2}}$ for all real values of $x > 2$. **(Derivative)**

19. $f(x) = \dfrac{3}{x - 2}$ for all real values of $x \neq 2$

$$f'(x) = \lim_{h \to 0} \frac{f(x + h) - f(x)}{h}$$

$$= \lim_{h \to 0} \frac{\dfrac{3}{(x + h) - 2} - \dfrac{3}{x - 2}}{h}$$

$$= \lim_{h \to 0} \frac{3(x - 2) - 3[(x + h) - 2]}{h[(x + h) - 2](x - 2)}$$

$$= \lim_{h \to 0} \frac{3x - 6 - 3x - 3h + 6}{h[(x + h) - 2](x - 2)}$$

$$= \lim_{h \to 0} \frac{-3h}{h[(x + h) - 2](x - 2)}$$

$$= \lim_{h \to 0} \frac{-3}{[(x + h) - 2](x - 2)} \qquad \text{(since } h \neq 0\text{)}$$

$$= \frac{-3}{(x - 2)(x - 2)}$$

$$= \frac{-3}{(x - 2)^2} \quad \text{for all real values of } x \neq 2$$

Therefore, if $f(x) = \dfrac{3}{x - 2}$, then $f'(x) = \dfrac{-3}{(x-2)^2}$ for all real values of $x \neq 2$. **(Derivative)**

20. $f(x) = x^3 - 1$ for all real values of x

$$f'(x) = \lim_{h \to 0} \frac{f(x + h) - f(x)}{h}$$

$$= \lim_{h \to 0} \frac{[(x + h)^3 - 1] - (x^3 - 1)}{h}$$

$$= \lim_{h \to 0} \frac{x^3 + 3hx^2 + 3h^2x + h^3 - 1 - x^3 + 1}{h}$$

$$= \lim_{h \to 0} \frac{3hx^2 + 3h^2x + h^3}{h}$$

$$= \lim_{h \to 0} [3x^2 + 3hx + h^2] \qquad \text{(since } h \neq 0\text{)}$$

$$= 3x^2 + 3x(0) + (0)^2$$

$$= 3x^2 \quad \text{for all real values of } x.$$

Therefore, if $f(x) = x^3 - 1$, then $f'(x) = 3x^2$ for all real values of x. **(Derivative)**

21. $f(x) = x^2 + x + 1$ for all real values of x

$$f'(x) = \lim_{h \to 0} \frac{f(x + h) - f(x)}{h}$$

$$= \lim_{h \to 0} \frac{[(x + h)^2 + (x + h) + 1] - (x^2 + x + 1)}{h}$$

$$= \lim_{h \to 0} \frac{x^2 + 2hx + h^2 + x + h + 1 - x^2 - x - 1}{h}$$

$$= \lim_{h \to 0} \frac{2hx + h^2 + h}{h}$$

$$= \lim_{h \to 0} (2x + h + 1) \quad \text{(since } h \neq 0\text{)}$$

$$= 2x + 0 + 1$$

$$= 2x + 1 \text{ for all real values of x.}$$

Therefore, if $f(x) = x^2 + x + 1$, then $f'(x) = 2x + 1$ for all real values of x. **(Derivative)**

22. $f(x) = 2x - x^3$ for all real values of x

$$f'(x) = \lim_{h \to 0} \frac{f(x + h) - f(x)}{h}$$

$$= \lim_{h \to 0} \frac{[2(x + h) - (x + h)^3] - (2x - x^3)}{h}$$

$$= \lim_{h \to 0} \frac{2x + 2h - [x^3 + 3hx^2 + 3h^2x + h^3] - 2x + x^3}{h}$$

$$= \lim_{h \to 0} \frac{2h - x^3 - 3hx^2 - 3h^2x - h^3 + x^3}{h}$$

$$= \lim_{h \to 0} \frac{2h - 3hx^2 - 3h^2x - h^3}{h}$$

$$= \lim_{h \to 0} [2 - 3x^2 - 3hx - h^2] \quad \text{(since } h \neq 0\text{)}$$

$$= 2 - 3x^2 - 3x(0) - (0)^2$$

$$= 2 - 3x^2 \text{ for all real values of x.}$$

Therefore, if $f(x) = 2x - x^3$, then $f'(x) = 2 - 3x^2$ for all real values of x. **(Derivative)**

23. $y = f(x) = 2x^2 + 1$ for all real values of x

$$m = f'(x) = \lim_{h \to 0} \frac{f(x + h) - f(x)}{h}$$

$$= \lim_{h \to 0} \frac{[2(x + h)^2 + 1] - [2x^2 + 1]}{h}$$

$$= \lim_{h \to 0} \frac{2[x^2 + 2hx + h^2] + 1 - 2x^2 - 1}{h}$$

$$= \lim_{h \to 0} \frac{4hx + 2h^2}{h}$$

$$= \lim_{h \to 0} (4x + 2h) \quad \text{(since } h \neq 0)$$

$$= 4x + 2(0)$$

$$= 4x \text{ for all real values of x.}$$

We will use the symbol $\underline{m}|_{x=1}$ to denote the slope of the tangent line when x = 1. Hence,

$$\underline{m}|_{x=1} = 4(1) = 4$$

Hence, the slope of the tangent line to the curve at the point where x = 1 is 4.

$$f(1) = 2(1)^2 + 1 = 2(1) + 1 = 3$$

Using the point–slope formula for the equation of a line with m = 4 and the point (1, 3), we have

$$y - y_1 = m(x - x_1)$$

$$y - 3 = 4(x - 1)$$

$$y - 3 = 4x - 4$$

$$y = 4x - 1, \text{ which is the required equation. \textbf{(Tangent line)}}$$

24. $y = f(x) = x^3 - 2$ for all real values of x

$$m = f'(x) = \lim_{h \to 0} \frac{f(x + h) - f(x)}{h}$$

$$= \lim_{h \to 0} \frac{[(x + h)^3 - 2] - (x^3 - 2)}{h}$$

$$= \lim_{h \to 0} \frac{[x^3 + 3hx^2 + 3h^2x + h^3 - 2] - (x^3 - 2)}{h}$$

$$= \lim_{h \to 0} \frac{3hx^2 + 3h^2x + h^3}{h}$$

$$= \lim_{h \to 0} [3x^2 + 3hx + h^2] \quad \text{(since } h \neq 0)$$

$$= 3x^2 + 3x(0) + (0)^2$$

$$= 3x^2 \text{ for all real values of x.}$$

$$\underline{m}|_{x=0} = 3(0)^2 = 0$$

Hence, the slope of the tangent line to the curve at the point where x = 0 is 0.

$$f(0) = (0)^3 - 2 = 0 - 2 = -2$$

Using the point–slope formula for the equation of a line with m = 0 and the point (0, –2), we have

$$y - y_1 = m(x - x_1)$$

$$y - (-2) = 0(x - 0)$$

$$y + 2 = 0$$

$$y = -2, \text{ which is the required equation. \textbf{(Tangent line)}}$$

25. y = f(x) = 8 for all real values of x

$$m = f'(x) = \lim_{h \to 0} \frac{f(x+h) - f(x)}{h}$$

$$= \lim_{h \to 0} \frac{8-8}{h}$$

$$= \lim_{h \to 0} \left(\frac{0}{h}\right)$$

$$= \lim_{h \to 0} (0) \qquad (\text{since } h \neq 0)$$

$$= 0 \text{ for all real values of x.}$$

$$m|_{x=3} = 0$$

Hence, the slope of the tangent line to the curve at the point where x = 3 is 0.

$$f(3) = 8$$

Using the point–slope formula for the equation of a line with m = 0 and the point (0, 8), we have

$$y - y_1 = m(x - x_1)$$

$$y - 8 = 0(x - 3)$$

$$y - 8 = 0$$

y = 8, which is the required equation. **(Tangent line)**

26. y = c for all real values of x

$$\frac{dy}{dx} = \frac{d(c)}{dx} = \lim_{h \to 0} \frac{c-c}{h}$$

$$= \lim_{h \to 0} \left(\frac{0}{h}\right)$$

$$= \lim_{h \to 0} (0) \qquad (\text{since } h \neq 0)$$

$$= 0 \text{ for all real values of x.}$$

Therefore, if y = c, then $\frac{dy}{dx} = 0$ for all real values of x.   **(Derivative)**

27. y = x for all real values of x

$$\frac{dy}{dx} = \frac{d(x)}{dx} = \lim_{h \to 0} \frac{(x+h)-x}{h}$$

$$= \lim_{h \to 0} \left(\frac{h}{h}\right)$$

$$= \lim_{h \to 0} (1) \qquad (\text{since } h \neq 0)$$

$$= 1 \text{ for all real values of x.}$$

Therefore, if y = x, then $\frac{dy}{dx} = 1$ for all real values of x.   **(Derivative)**

28. $y = x^2$ for all real values of x

$$\frac{dy}{dx} = \frac{d(x^2)}{dx} = \lim_{h \to 0} \frac{(x+h)^2 - x^2}{h}$$

$$= \lim_{h \to 0} \frac{x^2 + 2hx + h^2 - x^2}{h}$$

$$= \lim_{h \to 0} \frac{2hx + h^2}{h}$$

$$= \lim_{h \to 0} (2x + h) \quad (\text{since } h \neq 0)$$

$$= 2x + 0$$

$$= 2x \text{ for all real values of x.}$$

Therefore, if $y = x^2$, then $\frac{dy}{dx} = 2x$ for all real values of x.   (**Derivative**)

29. $f(x) = 13$ for all real values of x

$f'(x) = \dfrac{d(13)}{dx} = 0$ for all real values of x. (Rule (i))   (**Derivative**)

30. $f(x) = 2x + 7$ for all real values of x

$$f'(x) = \frac{d}{dx}(2x+7) = \frac{d(2x)}{dx} + \frac{d(7)}{dx} \qquad \text{(Rule (v))}$$

$$= 2\frac{d(x)}{dx} + \frac{d(7)}{dx} \qquad \text{(Rule (iv))}$$

$$= 2(1) + \frac{d(7)}{dx} \qquad \text{(Rule (ii))}$$

$$= 2 + 0 \qquad \text{(Rule (i))}$$

$$= 2 \text{ for all real values of x. (\textbf{Derivative})}$$

31. $f(x) = x^7$ for all real values of x

$$f'(x) = \frac{d(x^7)}{dx} = 7x^{7-1} \qquad \text{(Rule (iii))}$$

$$= 7x^6 \text{ for all real values of x. (\textbf{Derivative})}$$

32. $f(x) = x^{-5}$ for all real values of $x \neq 0$

$$f'(x) = \frac{d(x^{-5})}{dx} = -5x^{-5-1} \qquad \text{(Rule (iii))}$$

$$= -5x^{-6} \text{ for all real values of } x \neq 0. \ (\textbf{Derivative})$$

33. $f(x) = -7x$ for all real values of x

$$f'(x) = \frac{d}{dx}(-7x) = -7\frac{d(x)}{dx} \quad \text{(Rule (iv))}$$

$$= -7(1) \ \text{(Rule (ii))}$$

$$= -7 \text{ for all real values of x. (\textbf{Derivative})}$$

34. $f(x) = x^2 + 2x + \sqrt{x}$ for all real values of $x \geq 0$

$$f'(x) = \frac{d(x^2 + 2x + \sqrt{x})}{dx}$$

$$= \frac{d(x^2)}{dx} + \frac{d(2x)}{dx} + \frac{d(x)^{1/2}}{dx} \qquad \text{(Rule (v) extended)}$$

$$= 2x + 2 + \frac{1}{2}x^{-\frac{1}{2}}$$

or, $\quad = 2x + 2 + \dfrac{1}{2x^{\frac{1}{2}}}$ for all real values of $x > 0$. (**Derivative**)

35. $f(x) = (1 + 2x)(3 - x)$ for all real values of x

$$f'(x) = \frac{d}{dx}[(1 + 2x)(3 - x)]$$

$$= (1 + 2x)\frac{d}{dx}(3 - x) + (3 - x)\frac{d}{dx}(1 + 2x) \qquad \text{(Product Rule)}$$

$$= (1 + 2x)(-1) + (3 - x)(2)$$

$$= -1 - 2x + 6 - 2x$$

$$= 5 - 4x \text{ for all real values of } x. \text{ (}\textbf{Derivative of a product}\text{)}$$

36. $f(x) = (x^2 - 4)(3x + 8)$ for all real values of x

$$f'(x) = \frac{d}{dx}[(x^2 - 4)(3x + 8)]$$

$$= (x^2 - 4)\frac{d}{dx}(3x + 8) + (3x + 8)\frac{d}{dx}(x^2 - 4) \qquad \text{(Product Rule)}$$

$$= (x^2 - 4)(3) + (3x + 8)(2x)$$

$$= 3x^2 - 12 + 6x^2 + 16x$$

$$= 9x^2 + 16x - 12 \text{ for all real values of } x. \text{ (}\textbf{Derivative of a product}\text{)}$$

37. $f(x) = \sqrt[3]{x}(2 - x^4)$ for all real values of x

$$f'(x) = \frac{d}{dx}[\sqrt[3]{x}(2 - x^4)]$$

$$= \sqrt[3]{x}\frac{d}{dx}(2 - x^4) + (2 - x^4)\frac{d}{dx}(x)^{1/3} \qquad \text{(Product Rule)}$$

$$= \sqrt[3]{x}(-4x^3) + (2 - x^4)[\frac{1}{3}x^{-\frac{2}{3}}]$$

$$= -4x^3 \sqrt[3]{x} + \frac{1}{3}(2 - x^4)x^{-\frac{2}{3}} \text{ for all real values of } x. \qquad \text{(}\textbf{Derivative of a product}\text{)}$$

38. $f(x) = (4x^3 - 1)(x^2 + 7)$ for all real values of x

$$f'(x) = \frac{d}{dx}[(4x^3 - 1)(x^2 + 7)]$$

$$= (4x^3 - 1)\frac{d}{dx}(x^2 + 7) + (x^2 + 7)\frac{d}{dx}(4x^3 - 1) \qquad \text{(Product Rule)}$$

$$= (4x^3 - 1)(2x) + (x^2 + 7)(12x^2)$$

$$= 8x^4 - 2x + 12x^4 + 84x^2$$

$$= 20x^4 + 84x^2 - 2x \text{ for all real values of x. } \textbf{(Derivative of a product)}$$

39. $f(x) = -5x^7(4x^8 - 5)$ for all real values of x

$$f'(x) = \frac{d}{dx}[(-5x^7)(4x^8 - 5)]$$

$$= (-5x^7)\frac{d}{dx}(4x^8 - 5) + (4x^8 - 5)\frac{d}{dx}(-5x^7) \qquad \text{(Product Rule)}$$

$$= (-5x^7)(32x^7) + (4x^8 - 5)(-35)x^6$$

$$= -160x^{14} - 140x^{14} + 175x^6$$

$$= -300x^{14} + 175x^6 \text{ for all real values of x. } \textbf{(Derivative of a product)}$$

40. $f(x) = (2x^2 - 1)(3 - 5x^3)$ for all real values of x

$$f'(x) = \frac{d}{dx}[(2x^2 - 1)(3 - 5x^3)]$$

$$= (2x^2 - 1)\frac{d}{dx}(3 - 5x^3) + (3 - 5x^3)\frac{d}{dx}(2x^2 - 1) \qquad \text{(Product Rule)}$$

$$= (2x^2 - 1)(-15x^2) + (3 - 5x^3)(4x)$$

$$= -30x^4 + 15x^2 + 12x - 20x^4$$

$$= -50x^4 + 15x^2 + 12x \text{ for all real values of x. } \textbf{(Derivative of a product)}$$

41. $f(x) = (9 - \sqrt[5]{x})(x^6 + 3)$ for all real values of x

$$f'(x) = \frac{d}{dx}[(9 - \sqrt[5]{x})(x^6 + 3)]$$

$$= (9 - \sqrt[5]{x})\frac{d}{dx}(x^6 + 3) + (x^6 + 3)\frac{d}{dx}(9 - \sqrt[5]{x}) \qquad \text{(Product Rule)}$$

$$= (9 - \sqrt[5]{x})(6x^5) + (x^6 + 3)\left(\frac{-1}{5x^{4/5}}\right)$$

$$= 6x^5(9 - \sqrt[5]{x}) - \frac{x^6 + 3}{5x^{4/5}} \text{ for all real values of x} > 0. \qquad \textbf{(Derivative of a product)}$$

42. $f(x) = \dfrac{x+1}{x-2}$ for all real values of $x \neq 2$

$f'(x) = \dfrac{d}{dx}\left(\dfrac{x+1}{x-2}\right)$

$\quad = \dfrac{(x-2)\dfrac{d}{dx}(x+1) - (x+1)\dfrac{d}{dx}(x-2)}{(x-2)^2}$ (Quotient Rule)

$\quad = \dfrac{(x-2)(1) - (x+1)(1)}{(x-2)^2}$

$\quad = \dfrac{(x-2) - (x+1)}{(x-2)^2}$

$\quad = \dfrac{-3}{(x-2)^2}$ for all real values of $x \neq 2$. **(Derivative of a quotient)**

43. $f(x) = \dfrac{\sqrt{x}}{x+3}$ for all real values of $x \geq 0$

$f'(x) = \dfrac{d}{dx}\left(\dfrac{\sqrt{x}}{x+3}\right)$

$\quad = \dfrac{(x+3)\dfrac{d}{dx}(\sqrt{x}) - \sqrt{x}\dfrac{d}{dx}(x+3)}{(x+3)^2}$ (Quotient Rule)

$\quad = \dfrac{(x+3)(\frac{1}{2}x^{-1/2}) - \sqrt{x}(1)}{(x+3)^2}$

$\quad = \dfrac{\dfrac{x+3}{2\sqrt{x}} - \sqrt{x}}{(x+3)^2}$

$\quad = \dfrac{\dfrac{x+3}{2\sqrt{x}} - \dfrac{2x}{2\sqrt{x}}}{(x+3)^2}$

$\quad = \dfrac{x+3-2x}{2\sqrt{x}\,(x+3)^2}$

$\quad = \dfrac{3-x}{2\sqrt{x}\,(x+3)^2}$ for all real values of $x > 0$. **(Derivative of a quotient)**

44. $f(x) = \dfrac{4 - (1/x)}{3x^2}$ for all real values of $x \neq 0$

$f'(x) = \dfrac{d}{dx}\left(\dfrac{4 - (1/x)}{3x^2}\right)$

$$= \frac{3x^2 \frac{d}{dx}[4 - (\tfrac{1}{x})] - [4 - (\tfrac{1}{x})]\frac{d}{dx}(3x^2)}{(3x^2)^2} \qquad \text{(Quotient Rule)}$$

$$= \frac{3x^2(x^{-2}) - [4 - (\tfrac{1}{x})](6x)}{9x^4}$$

$$= \frac{3 - 6x[4 - (\tfrac{1}{x})]}{9x^4}$$

$$= \frac{3 - 24x + 6}{9x^4}$$

$$= \frac{9 - 24x}{9x^4} \text{ for all real values of } x \neq 0. \qquad \textbf{(Derivative of a quotient)}$$

45. $f(x) = \dfrac{2x - 3}{5 - 4x}$ for all real values of $x \neq 5/4$

$$f'(x) = \frac{d}{dx}\left(\frac{2x - 3}{5 - 4x}\right)$$

$$= \frac{(5 - 4x)\frac{d}{dx}(2x - 3) - (2x - 3)\frac{d}{dx}(5 - 4x)}{(5 - 4x)^2} \qquad \text{(Quotient Rule)}$$

$$= \frac{(5 - 4x)(2) - (2x - 3)(-4)}{(5 - 4x)^2}$$

$$= \frac{10 - 8x + 8x - 12}{(5 - 4x)^2}$$

$$= \frac{-2}{(5 - 4x)^2} \text{ for all real values of } x \neq 5/4. \qquad \textbf{(Derivative of a quotient)}$$

46. $f(x) = \dfrac{x - 3}{x^2 - 3x - 4}$ for all real values of $x \neq -1, 4$

$$f'(x) = \frac{d}{dx}\left(\frac{x - 3}{x^2 - 3x - 4}\right)$$

$$= \frac{(x^2 - 3x - 4)\frac{d}{dx}(x - 3) - (x - 3)\frac{d}{dx}(x^2 - 3x - 4)}{(x^2 - 3x - 4)^2} \qquad \text{(Quotient Rule)}$$

$$= \frac{(x^2 - 3x - 4)(1) - (x - 3)(2x - 3)}{(x^2 - 3x - 4)^2}$$

$$= \frac{x^2 - 3x - 4 - 2x^2 + 9x - 9}{(x^2 - 3x - 4)^2}$$

$$= \frac{-x^2 + 6x - 13}{(x^2 - 3x - 4)^2} \text{ for all } x \neq -1, 4. \qquad \textbf{(Derivative of a quotient)}$$

47. $f(x) = \dfrac{x^3 - 1}{x^4 + 1}$ for all real values of x

$$f'(x) = \frac{d}{dx}\left(\frac{x^3 - 1}{x^4 + 1}\right)$$

$$= \frac{(x^4 + 1)\dfrac{d}{dx}(x^3 - 1) - (x^3 - 1)\dfrac{d}{dx}(x^4 + 1)}{(x^4 + 1)^2} \qquad \text{(Quotient Rule)}$$

$$= \frac{(x^4 + 1)(3x^2) - (x^3 - 1)(4x^3)}{(x^4 + 1)^2}$$

$$= \frac{3x^6 + 3x^2 - 4x^6 + 4x^3}{(x^4 + 1)^2}$$

$$= \frac{-x^6 + 4x^3 + 3x^2}{(x^4 + 1)^2} \text{ for all real values of x.} \qquad \textbf{(Derivative of a quotient)}$$

48. $f(x) = \dfrac{3x - 1}{2 - \sqrt{x}}$ for all real values of $x \geq 0$ such that $x \neq 4$

$$f'(x) = \frac{d}{dx}\left(\frac{3x - 1}{2 - \sqrt{x}}\right)$$

$$= \frac{(2 - \sqrt{x})\dfrac{d}{dx}(3x - 1) - (3x - 1)\dfrac{d}{dx}(2 - \sqrt{x})}{(2 - \sqrt{x})^2} \qquad \text{(Quotient Rule)}$$

$$= \frac{(2 - \sqrt{x})(3) - (3x - 1)\left(\dfrac{-1}{2\sqrt{x}}\right)}{(2 - \sqrt{x})^2}$$

$$= \frac{\dfrac{2\sqrt{x}(2 - \sqrt{x})(3)}{2\sqrt{x}} - (3x - 1)\left(\dfrac{-1}{2\sqrt{x}}\right)}{(2 - \sqrt{x})^2}$$

$$= \frac{6\sqrt{x}(2 - \sqrt{x}) + (3x - 1)}{2\sqrt{x}(2 - \sqrt{x})^2}$$

$$= \frac{12\sqrt{x} - 6x + 3x - 1}{2\sqrt{x}(2 - \sqrt{x})^2}$$

$$= \frac{12\sqrt{x} - 3x - 1}{2\sqrt{x}(2 - \sqrt{x})^2} \text{ for all real values of } x > 0 \text{ and such that } x \neq 4 \qquad \textbf{(Derivative of a quotient)}$$

49. $f(x) = \dfrac{(2x - 3)(x + 1)}{x^2 + 1}$ for all real values of x

$$f'(x) = \frac{d}{dx}\left(\frac{(2x - 3)(x + 1)}{x^2 + 1}\right)$$

$$= \frac{(x^2 + 1)\dfrac{d}{dx}[(2x - 3)(x + 1)] - (2x - 3)(x + 1)\dfrac{d}{dx}(x^2 + 1)}{(x^2 + 1)^2}$$

$$= \frac{(x^2+1)[(2x-3)\frac{d}{dx}(x+1)+(x+1)\frac{d}{dx}(2x-3)]-(2x-3)(x+1)\frac{d}{dx}(x^2+1)}{(x^2+1)^2}$$

$$= \frac{(x^2+1)[(2x-3)(1)+(x+1)(2)]-(2x-3)(x+1)(2x)}{(x^2+1)^2}$$

$$= \frac{(x^2+1)(2x-3+2x+2)-(2x-3)(x+1)(2x)}{(x^2+1)^2}$$

$$= \frac{(x^2+1)(4x-1)-2x(2x^2-x-3)}{(x^2+1)^2}$$

$$= \frac{4x^3-x^2+4x-1-4x^3+2x^2+6x}{(x^2+1)^2}$$

$$= \frac{x^2+10x-1}{(x^2+1)^2} \text{ for all real values of x. } \textbf{(Derivative of a quotient)}$$

50. $f(x) = \dfrac{-3(2-5x^6)}{8}$ for all real values of x

$$f'(x) = \frac{d}{dx}\left(\frac{(-3(2-5x^6))}{8}\right) = \frac{-3}{8}\frac{d}{dx}(2-5x^6)$$

$$= \frac{-3}{8}(-5)(6x^5)$$

$$= \frac{45}{4}x^5 \text{ for all real values of x. } \textbf{(Derivative of a quotient)}$$

51. $f(x) = \dfrac{2x^3(3x-5)}{1-\sqrt{x}}$ for all real values of $x \geq 0$

$$f^1(x) = \frac{d}{dx}\left(\frac{2x^3(3x-5)}{1-\sqrt{x}}\right)$$

$$= \frac{(1-\sqrt{x})\frac{d}{dx}[2x^3(3x-5)]-(2x^3)(3x-5)\frac{d}{dx}(1-\sqrt{x})}{(1-\sqrt{x})^2}$$

$$= \frac{(1-\sqrt{x})\frac{d}{dx}[(2x^3(3x-5)]-(2x^3)(3x-5)(-\frac{1}{2}x^{-1/2})}{(1-\sqrt{x})^2}$$

$$= \frac{(1-\sqrt{x})[\frac{d}{dx}(6x^4-10x^3)]-(2x^3)(3x-5)(\frac{-1}{2}x^{-1/2})}{(1-\sqrt{x})^2}$$

$$= \frac{(1-\sqrt{x})(24x^3-30x^2)-(2x^3)(3x-5)(\frac{-1}{2}x^{-1/2})}{(1-\sqrt{x})^2}$$

$$= \frac{(1-\sqrt{x})(24x^3-30x^2)-(2x^3)(3x-5)(\frac{-1}{2\sqrt{x}})}{1-\sqrt{x})^2}$$

$$= \frac{2\sqrt{x}(1 - \sqrt{x})(24x^3 - 30x^2) + (2x^3)(3x - 5)}{2\sqrt{x}(1 - \sqrt{x})^2}$$

$$= \frac{(2\sqrt{x} - 2x)(24x^3 - 30x^2) + (2x^3)(3x - 5)}{2\sqrt{x}(1 - \sqrt{x})^2} \text{ for all real values of } x > 0. \textbf{ (Derivative of a quotient)}$$

52. $f(x) = (2x - 3)^4$ for all real values of x

$$f'(x) = \frac{d}{dx}(2x - 3)^4 = 4(2x - 3)^3 \frac{d}{dx}(2x - 3)$$

$$= 4(2x - 3)^3(2)$$

$$= 8(2x - 3)^3 \text{ for all real values of x. } \textbf{(General Power Rule)}$$

53. $f(x) = \sqrt{3 - 2x}$ for all real values of $x \le 3/2$

$$f'(x) = \frac{d}{dx}(\sqrt{3 - 2x}) = \frac{d}{dx}(3 - 2x)^{\frac{1}{2}}$$

$$= \frac{1}{2}(3 - 2x)^{-\frac{1}{2}}\frac{d}{dx}(3 - 2x)$$

$$= \frac{1}{2}(3 - 2x)^{-\frac{1}{2}}(-2)$$

$$= (-1)(3 - 2x)^{-\frac{1}{2}}$$

$$= \frac{-1}{\sqrt{3 - 2x}} \text{ for all real values of } x < 3/2. \textbf{ (General Power Rule)}$$

54. $f(x) = (x^3 + 1)^7$ for all real values of x

$$f'(x) = \frac{d}{dx}(x^3 + 1)^7 = 7(x^3 + 1)^6 \frac{d}{dx}(x^3 + 1)$$

$$= 7(x^3 + 1)^6(3x^2)$$

$$= 21x^2(x^3 + 1)^6 \text{ for all real values of x. } \textbf{(General Power Rule)}$$

55. $f(x) = (7x - 1)^{-3}$ for all real values of $x \ne 1/7$

$$f'(x) = \frac{d}{dx}(7x - 1)^{-3} = -3(7x - 1)^{-4}\frac{d}{dx}(7x - 1)$$

$$= -3(7x - 1)^{-4}(7)$$

$$= -21(7x - 1)^{-4} \text{ for all real values of } x \ne 1/7. \textbf{ (General Power Rule)}$$

56. $f(x) = (8x^6 + 4x - 1)^2$ for all real values of x

$$f'(x) = \frac{d}{dx}(8x^6 + 4x - 1)^2 = 2(8x^6 + 4x - 1)\frac{d}{dx}(8x^6 + 4x - 1)$$

$$= 2(8x^6 + 4x - 1)(48x^5 + 4)$$

$$= 8(12x^5 + 1)(8x^6 + 4x - 1) \text{ for all real values of x. } \textbf{(General Power Rule)}$$

57. $f(x) = (\sqrt[3]{3x - 5})$ for all real values of x

$$f'(x) = \frac{d}{dx}(\sqrt[3]{3x - 5}) = \frac{d}{dx}(3x - 5)^{1/3}$$

$$= \frac{1}{3}(3x - 5)^{-2/3}\frac{d}{dx}(3x - 5)$$

$$= \frac{1}{3}(3x - 5)^{-2/3}(3)$$

$$= (3x - 5)^{-2/3}$$

$$= \frac{1}{(3x - 5)^{2/3}} \text{ for all real values of x} \neq 5/3. \textbf{ (General Power Rule)}$$

58. $f(x) = (3x^5 - 4x^3 + x - 5)^3$ for all real values of x

$$f'(x) = \frac{d}{dx}(3x^5 - 4x^3 + x - 5)^3$$

$$= 3(3x^5 - 4x^3 + x - 5)^2\frac{d}{dx}(3x^5 - 4x^3 + x - 5)$$

$$= 3(3x^5 - 4x^3 + x - 5)^2[3(5)x^4 - 4(3)x^2 + 1 - 0]$$

$$= 3(3x^5 - 4x^3 + x - 5)^2(15x^4 - 12x^2 + 1) \text{ for all real values of x. } \textbf{(General Power Rule)}$$

59. $f(x) = (x^2 + 1)^3(2x - 1)$ for all real values of x

$$f'(x) = \frac{d}{dx}[(x^2 + 1)^3(2x - 1)]$$

$$= (x^2 + 1)^3\frac{d}{dx}(2x - 1) + (2x - 1)\frac{d}{dx}(x^2 + 1)^3$$

$$= (x^2 + 1)^3(2) + (2x - 1)[3(x^2 + 1)^2\frac{d}{dx}(x^2 + 1)]$$

$$= 2(x^2 + 1)^3 + (2x - 1)[3(x^2 + 1)^2(2x)]$$

$$= 2(x^2 + 1)^3 + 6x(2x - 1)(x^2 + 1)^2$$

$$= 2(x^2 + 1)^2[x^2 + 1 + 3x(2x - 1)]$$

$$= 2(x^2 + 1)^2(x^2 + 1 + 6x^2 - 3x)$$

$$= 2(x^2 + 1)^2(7x^2 - 3x + 1) \text{ for all real values of x. } \textbf{(Derivative of a product)}$$

60. $f(x) = \dfrac{2x(x - 3)}{x + 1}$ for all real values of x $\neq -1$

$$f'(x) = \frac{d}{dx}\left(\frac{2x(x - 3)}{x + 1}\right)$$

$$= \frac{(x + 1)\frac{d}{dx}[2x(x - 3)] - 2x(x - 3)\frac{d}{dx}(x + 1)}{(x + 1)^2}$$

$$= \frac{(x + 1)\frac{d}{dx}(2x^2 - 6x) - 2x(x - 3)\frac{d}{dx}(x + 1)}{(x + 1)^2}$$

$$= \frac{(x+1)[2(2)x-6] - 2x(x-3)(1)}{(x+1)^2}$$

$$= \frac{(x+1)(4x-6) - 2x(x-3)}{(x+1)^2}$$

$$= \frac{4x^2 - 6x + 4x - 6 - 2x^2 + 6x}{(x+1)^2}$$

$$= \frac{2x^2 + 4x - 6}{(x+1)^2} \text{for all real values of } x \neq -1. \quad \textbf{(Derivative of a quotient)}$$

61. $f(x) = f(x) = \dfrac{-3x(x+5)^3}{5\sqrt{x+2}}$ for all real values of $x > -2$

$$f'(x) = \frac{d}{dx}\left(\frac{-3x(x+5)^3}{5\sqrt{x+2}}\right)$$

$$= \frac{5\sqrt{x+2}\dfrac{d}{dx}[-3x(x+5)^3] - (-3x)(x+5)^3\dfrac{d}{dx}[5(x+2)^{\frac{1}{2}}]}{[5\sqrt{x+2}]^2}$$

$$= \frac{5\sqrt{x+2}[-3x\dfrac{d}{dx}(x+5)^3 + (x+5)^3\dfrac{d}{dx}(-3x)] + (3x)(x+5)^3\dfrac{d}{dx}[5(x+2)^{\frac{1}{2}}]}{25(x+2)}$$

$$= \frac{5\sqrt{x+2}[-3x(3)(x+5)^2\dfrac{d}{dx}(x+5) + (x+5)^3(-3)] + (3x)(x+5)^3\dfrac{d}{dx}[5(x+2)^{\frac{1}{2}}]}{25(x+2)}$$

$$= \frac{5\sqrt{x+2}[-3x(3)(x+5)^2(1) + (x+5)^3(-3)] + (3x)(x+5)^3\dfrac{d}{dx}[5(x+2)^{\frac{1}{2}}]}{25(x+2)}$$

$$= \frac{5\sqrt{x+2}[-3x(3)(x+5)^2 + (x+5)^3(-3)] + (3x)(x+5)^3(5)\dfrac{1}{2}(x+2)^{-\frac{1}{2}}\dfrac{d}{dx}(x+2)]}{25(x+2)}$$

$$= \frac{5\sqrt{x+2}[-9x(x+5)^2 + (-3)(x+5)^3] + (3x)(x+5)^3(5)\dfrac{1}{2}(x+2)^{-\frac{1}{2}}(1)]}{25(x+2)}$$

$$= \frac{5\sqrt{x+2}[-9x(x+5)^2 + (-3)(x+5)^3] + (\dfrac{15}{2}x)(x+5)^3(x+2)^{-\frac{1}{2}}]}{25(x+2)}$$

$$= \frac{5\sqrt{x+2}[-3(x+5)^2(3x+x+5)] + (\dfrac{15}{2}x)(x+5)^3(x+2)^{-\frac{1}{2}}]}{25(x+2)}$$

$$= \frac{5\sqrt{x+2}[-3(x+5)^2(4x+5)] + \dfrac{(15x)(x+5)^3]}{2\sqrt{x+2}}}{25(x+2)}$$

$$= \frac{\dfrac{10(x+2)[-3(x+5)^2(4x+5)]}{2\sqrt{x+2}} + \dfrac{(15x)(x+5)^3}{2\sqrt{x+2}}]}{25(x+2)}$$

$$= \frac{-30(x+2)(x+5)^2(4x+5) + (15x)(x+5)^3}{50(x+2)^{3/2}}$$

$$= \frac{-15(x+5)^2[2(x+2)(4x+5) - x(x+5)]}{50(x+2)^{3/2}}$$

$$= \frac{-15(x+5)^2[2(4x^2+13x+10) - x^2 - 5x]}{50(x+2)^{3/2}}$$

$$= \frac{-15(x+5)^2(8x^2+26x+20 - x^2 - 5x)}{50(x+2)^{3/2}}$$

$$= \frac{-3(x+5)^2(7x^2+21x+20)}{10(x+2)^{3/2}} \text{ for all real values of } x > -2. \textbf{ (Derivative of a quotient)}$$

62. $f(x) = x^3\sqrt{3x^2+5}$ for all real values of x

$$f'(x) = \frac{d}{dx}[x^3\sqrt{3x^2+5}]$$

$$= x^3\frac{d}{dx}\sqrt{3x^2+5} + \sqrt{3x^2+5}\frac{d}{dx}(x^3)$$

$$= x^3\left(\frac{1}{2}(3x^2+5)^{-1/2}\frac{d}{dx}(3x^2+5)\right) + \sqrt{3x^2+5}\,(3x^2)$$

$$= x^3\left(\frac{1}{2}(3x^2+5)^{-1/2}(6x)\right) + (3x^2)\sqrt{3x^2+5}$$

$$= 3x^4(3x^2+5)^{-1/2} + 3x^2\sqrt{3x^2+5}$$

$$= \frac{3x^4}{(3x^2+5)^{1/2}} + 3x^2\sqrt{3x^2+5}$$

$$= \frac{3x^4 + 3x^2(3x^2+5)}{\sqrt{3x^2+5}}$$

$$= \frac{3x^4 + 9x^4 + 15x^2}{\sqrt{3x^2+5}}$$

$$= \frac{12x^4 + 15x^2}{\sqrt{3x^2+5}}$$

$$= \frac{3x^2(4x^2+5)}{\sqrt{3x^2+5}} \text{ for all real values of x. } \textbf{(Derivative of a product)}$$

63. $f(x) = \dfrac{\sqrt{x}+3}{(x^2+2)^3}$ for all real values of $x \geq 0$

$$f'(x) = \frac{d}{dx}\left(\frac{\sqrt{x}+3}{(x^2+2)^3}\right)$$

$$= \frac{(x^2+2)^3 \frac{d}{dx}(\sqrt{x}+3) - (\sqrt{x}+3)\frac{d}{dx}(x^2+2)^3}{(x^2+2)^6}$$

$$= \frac{(x^2+2)^3(\frac{1}{2}x^{-\frac{1}{2}}) - (\sqrt{x}+3)(3)(x^2+2)^2\frac{d}{dx}(x^2+2)}{(x^2+2)^6}$$

$$= \frac{(x^2+2)^3(\frac{1}{2}x^{-\frac{1}{2}}) - (\sqrt{x}+3)(3)(x^2+2)^2(2x)}{(x^2+2)^6}$$

$$= \frac{(x^2+2)^3(\frac{1}{2}x^{-\frac{1}{2}}) - (\sqrt{x}+3)(6x)(x^2+2)^2}{(x^2+2)^6}$$

$$= \frac{\frac{(x^2+2)^3}{2\sqrt{x}} - (\sqrt{x}+3)(6x)(x^2+2)^2}{(x^2+2)^6}$$

$$= \frac{\frac{(x^2+2)^3}{2\sqrt{x}} - \frac{2\sqrt{x}(\sqrt{x}+3)(6x)(x^2+2)^2}{2\sqrt{x}}}{(x^2+2)^6}$$

$$= \frac{(x^2+2)^3 - 2\sqrt{x}(\sqrt{x}+3)(6x)(x^2+2)^2}{2\sqrt{x}(x^2+2)^6}$$

$$= \frac{(x^2+2)^3 - 12x(x+3\sqrt{x})(x^2+2)^2}{2\sqrt{x}(x^2+2)^6}$$

$$= \frac{(x^2+2)^2[x^2+2 - 12x^2 - 36x^{\frac{3}{2}}]}{2\sqrt{x}(x^2+2)^6}$$

$$= \frac{(x^2+2)^2[-11x^2 + 2 - 36x^{\frac{3}{2}}]}{2\sqrt{x}(x^2+2)^6} \text{ for all real values of } x > 0. \textbf{ (Derivative of a quotient)}$$

64. $f(x) = \dfrac{-5x^3}{\sqrt{x^2+4x+4}}$ for all real values of $x \neq -2$

$$f'(x) = \frac{d}{dx}\left(\frac{-5x^3}{\sqrt{x^2+4x+4}} \right)$$

$$= \frac{\sqrt{x^2+4x+4}\frac{d}{dx}(-5x^3) - (-5x^3)\frac{d}{dx}(\sqrt{x^2+4x+4})}{[(x^2+4x+4)^{\frac{1}{2}}]^2}$$

$$= \frac{\sqrt{x^2+4x+4}(-15x^2) - (-5x^3)\frac{1}{2}(x^2+4x+4)^{-\frac{1}{2}}\frac{d}{dx}(x^2+4x+4)}{x^2+4x+4}$$

$$= \frac{\sqrt{x^2+4x+4}(-15x^2) + (5x^3)\frac{1}{2}(x^2+4x+4)^{-\frac{1}{2}}(2x+4)}{x^2+4x+4}$$

$$= \frac{\sqrt{x^2+4x+4}(-15x^2) + \dfrac{(5x^3)(2x+4)}{2(x^2+4x+4)^{\frac{1}{2}}}}{x^2+4x+4}$$

$$= \frac{2(x^2+4x+4)(-15x^2) + (5x^3)(2x+4)}{2(x^2+4x+4)^{\frac{3}{2}}}$$

$$= \frac{(-5x^2)[2(x^2+4x+4)(3) - x(2x+4)]}{2(x^2+4x+4)^{\frac{3}{2}}}$$

$$= \frac{(-5x^2)[6x^2+24x+24 - 2x^2 - 4x]}{2(x^2+4x+4)^{\frac{3}{2}}}$$

$$= \frac{(-5x^2)[4x^2+20x+24]}{2(x^2+4x+4)^{\frac{3}{2}}}$$

$$= \frac{(-5x^2)[2x^2+10x+12]}{(x^2+4x+4)^{\frac{3}{2}}}$$

$$= \frac{-10x^4 - 50x^3 - 60x^2}{(x^2+4x+4)^{\frac{3}{2}}} \text{ for all real values of } x \neq -2. \quad \textbf{(Derivative of a quotient)}$$

65. $f(x) = \sqrt[3]{(1-x)^4}\,(x^2-3)^2$ for all real values of x

$$f'(x) = \frac{d}{dx}[\sqrt[3]{(1-x)^4}\,(x^2-3)^2]$$

$$= \sqrt[3]{(1-x)^4}\frac{d}{dx}(x^2-3)^2 + (x^2-3)^2\frac{d}{dx}(\sqrt[3]{(1-x)^4})$$

$$= \sqrt[3]{(1-x)^4}(2)(x^2-3)^1\frac{d}{dx}(x^2-3) + (x^2-3)^2\frac{d}{dx}(\sqrt[3]{(1-x)^4})$$

$$= \sqrt[3]{(1-x)^4}(2)(x^2-3)(2x) + (x^2-3)^2\frac{d}{dx}(\sqrt[3]{(1-x)^4})$$

$$= 4x(x^2-3)\sqrt[3]{(1-x)^4} + (x^2-3)^2\left(\frac{4}{3}(1-x)^{\frac{1}{3}}\frac{d}{dx}(1-x)\right)$$

$$= 4x(x^2-3)\sqrt[3]{(1-x)^4} + (x^2-3)^2\left(\frac{4}{3}(1-x)^{\frac{1}{3}}(-1)\right)$$

$$= 4x(x^2-3)\sqrt[3]{(1-x)^4} + (x^2-3)^2\left(\frac{-4}{3}(1-x)^{\frac{1}{3}}\right)$$

$$= \frac{-4}{3}(x^2-3)(1-x)^{\frac{1}{3}}[-3x(1-x) + (x^2-3)]$$

$$= \frac{-4}{3}(x^2-3)(1-x)^{\frac{1}{3}}(-3x+3x^2+x^2-3)$$

$$= \frac{-4}{3}(x^2-3)(1-x)^{\frac{1}{3}}(4x^2-3x-3) \text{ for all real values of x. } \textbf{(Derivative of a product)}$$

Grade Yourself

Circle the numbers of the questions you missed, then fill in the total incorrect for each topic. If you answered more than three questions incorrectly, you need to focus on that topic. (If a topic has less than three questions and you had at least one wrong, we suggest you study that topic also. Read your textbook, a review book, or ask your teacher for help.)

Subject: The Derivative

Topic	Question Numbers	Number Incorrect
Difference quotient	1, 2, 3, 4, 5,6	
Slope of a secant line	7, 8, 9, 10, 11, 12	
Average rate of change	13, 14, 15	
Derivative	16, 17, 18, 19, 20, 21, 22, 26, 27, 28, 29, 30, 31, 32, 33, 34	
Tangent line	23, 24, 25	
Derivative of a product	35, 36, 37, 38, 39, 40, 41, 59, 62, 65	
Derivative of a quotient	42, 43, 44, 45, 46, 47, 48, 49, 50, 51, 60, 61, 63, 64	
General Power Rule	52, 53, 54, 55, 56, 57, 58	

Marginal Analysis

4

 ## Test Yourself

4.1 Cost Analysis

Manufacturing companies produce items for sale in the hope of making a profit. There are certain costs involved. We will use the following notation:

x: The number of units produced in some time interval.

C(x):The **cost function** indicating what the company's costs are for producing x units. The function value, C(0) indicates the **fixed costs** (utilities, insurance, depreciation, etc.). In general, we have

C(x) = (Fixed Costs) + (Variable Costs)

where the **variable costs** depend upon the number of units produced.

$\overline{C}(x)$: The **average cost** per unit produced.

$$\overline{C}(x) = \frac{C(x)}{x} \text{ where } x > 0.$$

C'(x): **Marginal cost**, which is the instantaneous rate of change of C(x) with respect to x. C'(x) approximates the change in cost of producing one more unit at a production level of x units.

$\overline{C}'(x)$: **Marginal average cost**, which is the instantaneous rate of change of $\overline{C}(x)$ with respect to x. C'(x) approximates the change in average cost per unit that results from producing one more unit at a production level of x units.

1. Given the cost function C(x) = 25x + 3000 (in dollars), determine:

   a. the fixed costs.

   b. the variable costs.

   c. the cost of producing 25 units.

   d. the cost of producing the 26th unit.

   e. the cost of producing each unit.

2. The total cost per day, C(x) (in dollars), for producing x transistor radios is given by

   $$C(x) = 12 + 9x - \frac{x^2}{10}, 0 \le x \le 10.$$

   a. Compute C(5).

   b. Compute C(9).

   c. Determine the cost of producing the 6th radio.

   d. Compute $\frac{C(8) - C(7)}{8 - 7}$ and interpret the results.

3. A company produces x units of a product for which the fixed costs are $76,000 and the variable cost is $11.80 per unit.

   a. Determine the cost function, C(x).

   b. Compute C(8).

   c. Compute C(29) – C(28) and interpret the results.

   d. Determine the exact cost of producing the 31st item.

4. The cost of producing x units of a product is given by C(x) = 0.9x + 6000 (in dollars).

   a. Compute C(20).

b. Determine the average cost function, $\bar{C}(x)$.

c. Compute $\bar{C}(5000)$.

d. Compute $\bar{C}(10{,}000)$.

e. Compute $\bar{C}(100{,}000)$.

f. What conclusion can you draw from parts (c) – (e) above?

5. In Exercise 4, assume that the production level x approaches infinity. What is the limiting average cost per unit of the item?

6. The total cost per day, C(x) (in tens of dollars), for producing x racing bicycles is given by

$$C(x) = 480 + 20x - \frac{x^2}{4}, \; 0 \le x \le 40$$

Determine:

a. the marginal cost at a production level of x bicycles.

b. C′(29) and interpret the results.

c. the actual cost of producing the 30th bicycle.

d. Compare the results found in (b) and (c) above.

7. Let $C(x) = 0.4x^2 + 5x + 650$ be the cost function, in dollars, for producing x units of a product.

a. Determine an expression for $\bar{C}(x)$.

b. Compute $\bar{C}(32)$.

c. Determine an expression for C′(x).

d. Compute C′(26) and interpret the results.

e. Determine an expression for $\bar{C}\,'(x)$.

f. Compute $\bar{C}'(50)$ and interpret the results.

4.2 Revenue Analysis

When manufacturing companies produce and sell items, revenue is generated. We will use the following notation:

x: The number of units produced and sold in some time interval.

R(x): The **revenue function** indicating how much money is received from the production and sale of x units of an item.

R(x) = xp where x is the **demand** (that is, the number of units that can be sold) at the price p.

$\bar{R}(x)$: The **average revenue** per unit produced and sold.

$$\bar{R}(x) = \frac{R(x)}{x} \text{ where } x > 0.$$

R′(x): **Marginal revenue**, which is the instantaneous rate of change of R(x) with respect to x. R′(x) approximates the change in revenue resulting in producing and selling one more unit at a production and sale level of x units.

$\bar{R}'(x)$: **Marginal average revenue**, which is the instantaneous rate of change of R(x) with respect to x that results from producing and selling one more unit at a production and sale level of x units.

8. The marketing research department of a manufacturing company determined that the demand for its product is x units at the price of p dollars per unit, given by the **demand equation** x = 9000 – 1000p.

a. Solve for p in terms of x.

b. Determine an expression for the revenue function, R(x).

c. Compute R(12).

d. Compute R(110).

9. The total revenue (in dollars) from the production and sale of x units of an item is given by

$$R(x) = 100x - \frac{x^2}{50}$$

a. Determine the demand equation as a function of x.

b. Compute R(100).

c. Compute R(200).

d. Determine an expression for $\bar{R}(x)$.

e. Compute $\bar{R}(200)$ and interpret the result.

10. The marketing research department for a manufacturing company determined that the relationship between price p (in dollars) and the demand x (in units per week) is given by

$$p = 1450 - 0.15x^2 , 0 < x \leq 100.$$

   a. Determine an expression for R(x).

   b. Compute R(46).

   c. Compute R(68).

   d. Determine the average revenue function, $\overline{R}(x)$.

   e. Compute $\overline{R}(60)$ and interpret the result.

   f. Compute $\overline{R}(85)$.

11. Let R(x) = x(30 – 0.2x) be the total revenue function (in dollars).

   a. Compute R(100).

   b. Determine an expression for R'(x).

   c. Compute R'(50) and interpret the result.

   d. Compute R'(100) and interpret the result.

12. Let p = –0.6x + 750 be a demand equation (p in dollars).

   a. Determine an expression for R(x).

   b. Determine an expression for $\overline{R}(x)$.

   c. Determine an expression for R'(x).

   d. Determine an expression for $\overline{R}'(x)$.

13. Let $p = \dfrac{1000}{x + 6}$ be a demand equation with p in dollars. Determine the marginal revenue function R'(x) and compute R'(55).

4.3 Profit Analysis

When products are produced and sold, companies expect to make a profit. We will use the following notation:

x: The number of units produced and sold in some time interval.

P(x): The **profit function** indicating what the company's total profits are for producing and selling x units.

$$P(x) = R(x) - C(x)$$

where C(x) is the cost function and R(x) is the revenue function.

$\overline{P}(x)$: The **average profit per unit** when x units are produced and sold.

P'(x): **Marginal profit**, which is the instantaneous rate of change of P(x) with respect to x. P'(x) approximates the change in profit resulting from producing and selling one additional unit at a production and sale level of x units.

$\overline{P}'(x)$: **Marginal average profit**, which is the instantaneous rate of change of $\overline{P}(x)$ with respect to x. $\overline{P}'(x)$ approximates the change in average profit per unit resulting from producing and selling one additional unit at a level of production and sale of x units.

14. The total profit (in dollars) from the sale of x units of a product is given by

$$P(x) = 40x - \frac{x^2}{5} + 200.$$

   a. Compute P(200).

   b. Compute P(300).

   c. Determine the exact profit from the production and sale of the 79th unit.

15. Let C(x) = 800 + 70x be the cost function (in dollars) for the production of x units of an item, and let $R(x) = 300x - \dfrac{x^2}{20}$ be the corresponding revenue function (in dollars).

   a. Determine an expression for P(x).

   b. Compute P(80).

   c. Compute P(140).

   d. Determine the exact profit from the production and sale of the 57th unit.

16. The total profit (in dollars) from the production and sale of x units of a product is given by

$$P(x) = 18x - \frac{x^2}{15} - 350.$$

  a. Compute P(90).

  b. Determine an expression for $\overline{P}(x)$.

  c. Compute $\overline{P}(70)$ and interpret the result.

  d. Compute $\overline{P}(100)$ and interpret the result.

17. Refer to the profit function in Exercise 16.

  a. Determine an expression for $\overline{P}(x)$.

  b. Compute $\overline{P}(92)$ and interpret the result.

  c. Compute $\overline{P}(161)$ and interpret the result.

  d. Determine the exact profit from the production and sale of the 161st unit.

Definition: **Break–even points** are the prices where the cost and the revenue are equal; that is, the profit is zero.

18. The total cost and the total revenue functions for the production and sale of x units of a product are given by C(x) = 12x + 11,000 and R(x) = 100x − 0.1x² (in dollars), respectively. Estimate the production and sale level when P(x) = 0 (in dollars).

19. If the average cost function and the average revenue functions are given by

$$\overline{C}(x) = x + 160 + \frac{2000}{x} \text{ and}$$

$$\overline{R}(x) = 250 + 45x - x^2 \text{ (in dollars), respectively,}$$

  determine an expression for P′(x).

20. If the total cost function and the total profit function are given by C(x) = 4x + 115 and P(x) = 16x − x² − 125 (in dollars), respectively, determine the demand equation.

 # Check Yourself

1. C(x) = 25x + 3000 (in dollars)

  a. The fixed costs = C(0) = $[25(0) + 3000] = $3000.

  b. The variable costs are $25x.

  c. C(25) = $25(25) + $3000

  $\qquad = $625 + 3000

  $\qquad = 3625

  d. The cost of producing the 26th unit is given by C(26) − C(25)

  $\qquad = $[25(26) + 3000] − $[25(25) + 3000]$

  $\qquad = $(650 + 3000) − $(625 + 3000)$

  $\qquad = $3650 − 3625

  $\qquad = 25

  e. The cost of producing each unit is $25, the cofficient of x in the *linear* equation for C(x). (**Cost function**)

2. Determine $C(x) = 12 + 9x - \dfrac{x^2}{10}$ (in dollars)

a. $C(5) = \$\left(12 + 9(5) - \dfrac{5^2}{10}\right)$

$= \$\left(12 + 45 - \dfrac{25}{10}\right)$

$= \$(57 - 2.50)$

$= \$54.50$

b. $C(9) = \$\left(12 + 9(9) - \dfrac{9^2}{10}\right)$

$= \$\left(12 + 81 - \dfrac{81}{10}\right)$

$= \$(93 - 8.10)$

$= \$84.90$

c. $C(6) - C(5)$

$= \$\left[12 + 9(6) - \dfrac{6^2}{10}\right] - \$\left[12 + 9(5) - \dfrac{5^2}{10}\right]$

$= \$\left[12 + 54 - \dfrac{36}{10}\right] - \$\left[12 + 45 - \dfrac{25}{10}\right]$

$= \$(66 - 3.60) - \$(57 - 2.50)$

$= \$(62.40 - 54.50)$

$= \$7.90$

d. $\dfrac{C(8) - C(7)}{8 - 7}$ (in dollars)

$= \dfrac{\$\left[12 + 9(8) - \dfrac{8^2}{10}\right] - \$\left[12 + 9(7) - \dfrac{7^2}{10}\right]}{1}$

$= \$\left[12 + 72 - \dfrac{64}{10}\right] - \$\left[12 + 63 - \dfrac{49}{10}\right]$

$= \$(84 - 6.40) - \$(75 - 4.90)$

$= \$(77.60 - 70.10)$

$= \$7.50$ which is the cost of producing the 8th transistor radio. (**Cost function**)

3. Determine

a. $C(x) = 76,000 + 11.8x$ (in dollars) with $x \geq 0$.

b. $C(8) = \$76,000 + \$(11.80)(8)$

$= \$76,000 + \$94.40$

$= \$76,094.40$

c. C(29) – C(28) (in dollars)

$$= \$[76,000 + (11.8)(29)] - \$[76,000 + (11.8)(28)]$$

$$= \$(76,000 + 342.20) - \$(76,000 + 330.40)$$

$$= \$(76,342.20 - 76,330.40)$$

$$= \$11.80 \quad \text{which is the exact cost of producing the 29th unit.}$$

d. C(31) – C(30) (in dollars)

$$= \$[76,000 + (11.8)(30)] - \$[76,000 + (11.8)(30)]$$

$$= \$(76,000 + 365.80) - \$(76,000 + 354)$$

$$= \$(76,365.80 - 76,354)$$

$$= \$11.80 \quad \textbf{(Cost function)}$$

Note: The exact cost of producing the 29th unit (part c.) is the same as the exact cost of producing the 31st unit (part d.). In this exercise, the cost function, C(x), is a linear function and $11.80 is the coefficient of x. Hence, $11.80 is always the exact cost of producing the (x+1)–st unit.

4. C(x) = 0.9x + 6000 (in dollars)

a. C(20) = $[(0.9)(20) + 6000]

$$= \$(18 + 6000)$$

$$= \$6018$$

b. $\bar{C}(x) = \dfrac{C(x)}{x} = \dfrac{0.9x + 6000}{x}$ (in dollars)

c. $\bar{C}(5000) = \dfrac{\$[(0.9)(5000) + 6000]}{5000}$

$$= \dfrac{(\$4500 + 6000)}{5000}$$

$$= \dfrac{\$10,500}{5000}$$

$$= \$2.10 \quad \text{which is the } \textit{average cost per unit} \text{ at a production level of 5000 units}$$

d. $\bar{C}(10,000) = \dfrac{\$[(0.9)(10,000) + 6000]}{10,000}$

$$= \dfrac{\$(9000 + 6000)}{10,000}$$

$$= \dfrac{\$15,000}{10,000}$$

$$= \$1.50 \quad \text{which is the } \textit{average cost per unit} \text{ at a production level of 10,000 units.}$$

e. $\bar{C}(100,000) = \dfrac{\$[(0.9)(100,000) + 6000]}{100,000}$

$$= \dfrac{\$(90,000 + 6000)}{100,000}$$

$$= \frac{\$96,000}{100,000}$$

= \$.96   which is the *average cost per unit* at a production level of 100,000 units.

f. As the production level increases, the average cost per unit decreases. (**Average cost Function**)

Note: In this exercise, the cost function, C(x), is a linear function. The production cost per unit is always the same, \$.90 (the coefficient of x), but the fixed cost *per unit* decreases as more units are produced.

5. $\overline{C}(x) = \frac{0.9x + 6000}{x}$ (in dollars)

$$= \$\left(0.9 + \frac{6000}{x}\right)$$

$$\lim_{x \to \infty} \overline{C}(x) = \lim_{x \to \infty} \$(0.9 + \frac{6000}{x}) = \$.90$$

Hence, the limiting average cost per unit is \$.90 . (**Average cost Function**)

Note: Again, in this exercise, the cost function, C(x), is a linear function. The production cost per unit is always the same, \$.90 (the coefficient of x). We note above that the limiting average cost per unit is also \$.90. Hence, as the production level increases, the fixed cost *per unit* approaches \$0.

6. $C(x) = 480 + 20x - \frac{x^2}{4}$

a. $C'(x) = \frac{d}{dx}(C(x)) = \frac{d}{dx} \$(480 + 20x - \frac{x^2}{4})$

$$= \$(0 + 20 - \frac{2x}{4})$$

$$= \$(20 - \frac{x}{2})$$

b. $C'(29) = 20 - \frac{(2)(29)}{4}$ (in dollars)

$$= \$(20 - \frac{58}{4})$$

$$= \$(20 - 14.50)$$

= \$5.50 which is the approximate cost of producing the 30th unit.

c. C(30) − C(29)

$$= \$[480 + (20)(30) - \frac{30^2}{4}] - \$[480 + (20)(29) - \frac{29^2}{4}]$$

$$= \$[480 + 600 - \frac{900}{4}] - \$[480 + 580 - \frac{841}{4}]$$

$$= \$(1080 - 225) - \$(1060 - 210.25)$$

$$= \$(855 - 849.75)$$

$$= \$5.25$$

d. The actual cost of producing the 30th unit is $5.25 [from (c) above] whereas the approxmate cost of producing the 30th unit is $5.50 [from (b) above]. Normally, it is easier to compute the approximate cost (that is, the marginal cost) than it is to compute the actual cost of producing the next item. Oftentimes, the approximate cost is sufficient for making marketing decisions. (**Marginal cost Function**)

7. $C(x) = 0.4x^2 + 5x + 650$ (in dollars)

a. $\bar{C}(x) = \dfrac{C(x)}{x} = \dfrac{0.4x^2 + 5x + 650}{x}$ (in dollars)

$$= 0.4x + 5 + \dfrac{650}{x} \text{ (in dollars)}$$

b. $\bar{C}(32) = \$[(0.4)(32) + 5 + \dfrac{650}{32}]$

$$= \$[12.8 + 5 + 20\dfrac{5}{16}]$$

$$= \$38\dfrac{9}{80}$$

c. $C'(x) = \dfrac{d}{dx}(C(x)) = \dfrac{d}{dx}(0.4x^2 + 5x + 650)$ (in dollars)

$$= 0.8x + 5 + 0$$

$$= 0.8x + 5 \text{ (in dollars)}$$

d. $C'(26) = \$[(0.8)(26) + 5]$

$$= \$(20.80 + 5)$$

$$= \$25.80 \text{ which is the approximate cost of producing the 27th unit.}$$

e. $\bar{C}'(x) = \dfrac{d}{dx}(\bar{C}(x)) = \dfrac{d}{dx}(0.4x + 5 + \dfrac{650}{x})$ (in dollars)

$$= 0.4 + 0 - \dfrac{650}{x^2}$$

$$= 0.4 - \dfrac{650}{x^2} \text{ (in dollars)}$$

f. $\bar{C}'(50) = \$(0.4 - \dfrac{650}{(50)^2})$

$$= \$(0.4 - \dfrac{650}{2500})$$

$$= \$(0.4 - 0.26)$$

$$= \$.14 \text{ which is the approximate average cost per unit at a production level of 50 units.}$$
(**Marginal cost function**)

8. Demand equation: $x = 9000 - 1000p$ (p in dollars)

a. $x = 9000 - 1000p$

$p = \dfrac{9000 - x}{1000}$ (in dollars)

b. $R(x) = xp$ (in dollars)

$$= x\left(\dfrac{9000 - x}{1000}\right)$$

$$= \dfrac{9000x - x^2}{1000}$$ (in dollars)

c. $R(12) = \dfrac{(9000)(12) - (12)^2}{1000}$ (in dollars)

$$= \dfrac{108,000 - 144}{1000}$$

$$= \dfrac{107,856}{1000}$$

$$= \$107.86$$

d. $R(110) = \$\left(\dfrac{(9000)(110) - (110)^2}{1000}\right)$

$$= \$\left(\dfrac{990,000 - 12,100}{1000}\right)$$

$$= \$\left(\dfrac{977,900}{1000}\right)$$

$$= \$977.90 \text{ (\textbf{Revenue function})}$$

9. $R(x) = 100x - \dfrac{x^2}{50}$ (in dollars)

a. $R(x) = xp$

$$100x - \dfrac{x^2}{50} = xp$$

$$100 - \dfrac{x}{50} = p$$

$$5000 - x = 50p$$

$$x = 5000 - 50p \text{ (in dollars)}$$

b. $R(100) = \$[(100)(100) - \dfrac{(100)^2}{50}]$

$$= \$(10,000 - \dfrac{10,000}{50})$$

$$= \$(10,000 - 200)$$

$$= \$9800$$

c. $R(200) = \$[(100)(200) - \frac{(200)^2}{50}]$

$= \$(20,000 - \frac{40,000}{50})$

$= \$(20,000 - 800)$

$= \$19,200$

d. $\bar{R}(x) = \frac{R(x)}{x} = (100x - \frac{x^2}{50}) \div x$ (in dollars)

$= 100 - \frac{x}{50}$ (in dollars)

e. $\bar{R}(200) = \$(100 - \frac{200}{50})$

$= \$(100 - 4)$

$= \$96$ which is the *average revenue per unit* at a production and sale level of 200 units.
(Average revenue function)

10. Demand equation: $p = 1450 - 0.15x^2$ (p in dollars)

   a. $R(x) = xp$ (in dollars)

   $= x(1450 - 0.15x^2)$ (in dollars)

   $= 1450x - 0.15x^3$ (in dollars)

   b. $R(46) = \$[(1450)(46) - (0.15)(46)^3]$

   $= \$[66,700 - (0.15)(97,336)]$

   $= \$(66,700 - 14,600.40)$

   $= \$52,099.60$

   c. $\bar{R}(68) = \$[(1450)(68) - (0.15)(68)^3]$

   $= \$[98,600 - (0.15)(314,432)]$

   $= \$(98,600 - 47,164.80)$

   $= \$51,435.20$

   d. $\bar{R}(x) = \frac{R(x)}{x} = \frac{1450x - 0.15x^3}{x}$ (in dollars)

   $= 1450 - 0.15x^2$ (in dollars)

   e. $\bar{R}(60) = \$[1450 - (0.15)(60)^2]$

   $= \$[1450 - (0.15)(3600)]$

   $= \$(1450 - 540)$

   $= \$910$ which is the *average revenue per unit* at a production and sale level of 60 units.

f. $\overline{R}(85) = \$[1450 - (0.15)(85)^2]$

$\qquad = \$[1450 - (0.15)(7225)]$

$\qquad = \$(1450 - 1083.75)$

$\qquad = \$366.25$ which is the average revenue per unit at a production and sale level of 85 units. (**Average revenue function**)

11. $R(x) = x(30 - 0.2x)$ (in dollars)

a. $R(100) = 100[30 - (0.2)(100)]$ (in dollars)

$\qquad = 100(30 - 20)$

$\qquad = (100)(10)$ (in dollars)

$\qquad = \$1000$

b. $R'(x) = \dfrac{d}{dx}[x(30 - 0.2x)]$ (in dollars)

$\qquad = \dfrac{d}{dx}(30x - 0.2x^2)$

$\qquad = 30 - 0.4x$ (in dollars)

c. $R'(50) = \$[30 - (0.4)(50)]$

$\qquad = \$(30 - 20)$

$\qquad = \$10$ which approximates the revenue received from the production and sale of one additional unit after the production and sale of the 50th unit.

d. $R'(100) = \$[30 - (0.4)(100)]$

$\qquad = \$(30 - 40)$

$\qquad = -\$10$ (a loss) which approximates the revenue received from the production and sale of one additional unit after the production and sale of the 100th unit. (**Marginal revenue function**)

12. $p = -0.6x + 750$ (in dollars)

a. $R(x) = xp$ (in dollars)

$\qquad = x(-0.6x + 750)$

$\qquad = -0.6x^2 + 750x$ (in dollars)

b. $\overline{R}(x) = \dfrac{R(x)}{x} = \dfrac{-0.6x^2 + 750x}{x}$ (in dollars)

$\qquad = -0.6x + 750$ (in dollars)

c. $R'(x) = \dfrac{d}{dx}(R(x)) = \dfrac{d}{dx}(-0.6x^2 + 750x)$ (in dollars)

$\qquad = -1.2x + 750$ (in dollars)

d. $\overline{R}'(x) = \dfrac{d}{dx}(\overline{R}(x)) = \dfrac{d}{dx}(-0.6x + 750)$ (in dollars)

$\qquad = -0.6 + 0$ (in dollars)

$\qquad = -\$.60$ (**Marginal revenue function**)

13. $p = \dfrac{1000}{x+6}$ (p in dollars)

$R(x) = xp$ (in dollars)

$$= x(\dfrac{1000}{x+6})$$

$$= \dfrac{1000x}{x+6} \text{ (in dollars)}$$

$$R\,'(x) = \dfrac{d}{dx}(R(x)) = \dfrac{d}{dx}\left(\dfrac{1000x}{x+6}\right) \text{(in dollars)}$$

$$= \dfrac{(x+6)\dfrac{d}{dx}(1000x) - 1000x\dfrac{d}{dx}(x+6)}{(x+6)^2}$$

$$= \dfrac{(x+6)(1000) - (1000)x(1+0)}{(x+6)^2}$$

$$= \dfrac{1000[x+6-x]}{(x+6)^2}$$

$$= \dfrac{6000}{(x+6)^2} \text{ (in dollars)}$$

$$R\,'(55) = \$\left[\dfrac{6000}{(55+6)^2}\right]$$

$$= \$(\dfrac{6000}{(61)^2})$$

$$= \$(\dfrac{6000}{3721})$$

$\approx \$1.61$ which means that producing and selling one more unit after the 55th unit will result in approximately \$1.61 additional revenue. (**Marginal revenue function**)

14. $P(x) = 40x - \dfrac{x^2}{5} + 200$ (in dollars)

a. $P(200) = \$\left((40)(200) - \dfrac{(200)^2}{5} + 200\right)$

$$= \$(8000 - \dfrac{40,000}{5} + 200)$$

$$= \$(8000 - 8000 + 200)$$

$$= \$200$$

b. $P(300) = \$\left((40)(300) - \dfrac{(300)^2}{5} + 200\right)$

$$= \$\left(12,000 - \dfrac{90,000}{5} + 200\right)$$

$$= \$(12,000 - 18,000 + 200)$$

$$= -\$5800 \text{ (a loss)}$$

c. $P(79) - P(78)$ (in dollars)

$$= \$[(40)(79) - \frac{(79)^2}{5} + 200] - \$[(40)(78) - \frac{(78)^2}{5} + 200]$$

$$= \$[3160 - \frac{6241}{5} + 200] - \$[3120 - \frac{6084}{5} + 200]$$

$$= \$(3160 - 1248.20 + 200) - \$(3120 - 1216.80 + 200)$$

$$= \$(2111.80 - 2103.20)$$

$$= \$8.60 \text{ (\textbf{Profit function})}$$

15. $C(x) = 800 + 70x$ and $R(x) = 300x - \frac{x^2}{20}$ (in dollars)

a. $P(x) = R(x) - C(x)$ (in dollars)

$$= [300x - \frac{x^2}{20}] - (800 + 70x)$$

$$= 300x - \frac{x^2}{20} - 800 - 70x$$

$$= 230x - \frac{x^2}{20} - 800 \text{ (in dollars)}$$

b. $P(80) = \$\left((230)(80) - \frac{(80)^2}{20} - 800\right)$

$$= \$(18,400 - 320 - 800)$$

$$= \$17,280$$

c. $P(140) = \$[(230)(140) - \frac{(140)^2}{20} - 800]$

$$= \$(32,200 - 980 - 800)$$

$$= \$30,420$$

d. $P(57) - P(56)$ (in dollars)

$$= \$[(230)(57) - \frac{(57)^2}{20} - 800] - \$[(230)(56) - \frac{(56)^2}{20} - 800]$$

$$= \$[13,110 - \frac{3249}{20} - 800] - \$[12,880 - \frac{3136}{20} - 800]$$

$$= \$(13,110 - 162.45 - 800) - \$(12,880 - 156.80 - 800)$$

$$= \$(12,147.55 - 11,923.20)$$

$$= \$224.35 \text{ (\textbf{Profit function})}$$

16. $P(x) = 18x - \dfrac{x^2}{15} - 350$ (in dollars)

   a. $P(90) = \$[(18)(90) - \dfrac{(90)^2}{15} - 350]$

$$= \$(1620 - \dfrac{8100}{15} - 350)$$

$$= \$(1620 - 540 - 350)$$

$$= \$730$$

   b. $\overline{P}(x) = \dfrac{P(x)}{x}$ (in dollars)

$$= [18x - \dfrac{x^2}{15} - 350] \div x$$

$$= 18 - \dfrac{x}{15} - \dfrac{350}{x}$$ (in dollars)

   c. $\overline{P}(70) = \$(18 - \dfrac{70}{15} - \dfrac{350}{70})$

$$= \$(18 - 4\tfrac{2}{3} - 5)$$

    $\approx \$8.33$ which is the *average profit per unit* at a production and sale level of 70 units.

   d. $\overline{P}(100) = \$(18 - \dfrac{100}{15} - \dfrac{350}{100})$

$$= \$(18 - 6\tfrac{2}{3} - 3.5)$$

    $\approx \$7.83$ which is the *average profit per unit* at a production and sale level of 100 units.
(**Average profit function**)

17. $P(x) = 18x - \dfrac{x^2}{15} - 350$ (in dollars)

   a. $\overline{P}(x) = 18 - \dfrac{x}{15} - \dfrac{350}{x}$ (in dollars) (from Example 16b.)

   b. $\overline{P}(92) = \$\left(18 - \dfrac{92}{15} - \dfrac{350}{92}\right)$

$$= \$(18 - 6\tfrac{2}{15} - 3\tfrac{37}{46})$$

    $\approx \$8.06$ which is the *average profit per unit* at a production and sale level of 92 units.

   c. $\overline{P}(161) = \$(18 - \dfrac{161}{15} - \dfrac{350}{161})$

$$= \$(18 - 10\tfrac{11}{15} - 2\tfrac{28}{161})$$

    $\approx \$5.09$ which is the *average profit per unit* at a production and sale level of 161 units.

d. $P(161) - P(160)$ (in dollars)

$$= \$[(18)(161) - \frac{(161)^2}{15} - 350] - \$[(18)(160) - \frac{(160)^2}{15} - 350]$$

$$= \$[2898 - \frac{25,921}{15} - 350] - \$[2880 - \frac{25,600}{15} - 350]$$

$$= \$(2898 - 1728\frac{1}{15} - 350) - \$(2880 - 1706\frac{2}{3} - 350)$$

$$= \$(819\frac{14}{15} - 823\frac{1}{3})$$

$$= -\$3.40 \ (\textbf{Average profit function})$$

18. $C(x) = 12x + 11,000$ and $R(x) = 100x - 0.1x^2$ (in dollars)

$P(x) = R(x) - C(x)$ (in dollars)

$$= (100x - 0.1x^2) - (12x + 11,000)$$

$$= 100x - 0.1x^2 - 12x - 11,000$$

$$= -0.1x^2 + 88x - 11,000 \text{ (in dollars)}$$

Set $P(x) = 0$ (that is, revenue = cost)

$$-0.1x^2 + 88x - 11,000 = 0$$

$$x^2 - 880x + 110,000 = 0$$

$$x = \frac{-(-880) \pm \sqrt{(-880)^2 - 4(1)(110,000)}}{(2)(1)}$$

$$x = \frac{880 \pm \sqrt{774,400 - 440,000}}{2}$$

$$x \doteq \frac{880 \pm \sqrt{334,400}}{2}$$

$$x \approx \frac{880 \pm 578.27}{2}$$

$$x \approx \frac{880 + 578.27}{2} \qquad \text{or} \qquad x \approx \frac{880 - 578.27}{2}$$

$$x \approx 729.14 \qquad\qquad\qquad x \approx 150.87$$

Hence, profit will be $0 when the number of units produced and sold is approximately 151 or 729. Hence, the break-even points are 151 and 729. These are the number of units of production and sale where the cost equals the revenue. (It can be shown that when $x < 151$ or when $x > 729$, a loss will occur. However, when $151 \leq x \leq 729$, a profit will occur.) (**Profit function**)

19. $\bar{C}(x) = x + 160 + \dfrac{2000}{x}$ and $\bar{R}(x) = 250 + 45x - x^2$ (in dollars)

Since $\bar{C}(x) = \dfrac{C(x)}{x}$, we have

$$C(x) = (x)\bar{C}(x)$$

$$= x(x + 160 + \dfrac{2000}{x})$$

$$= x^2 + 160x + 2000 \text{ (in dollars)}$$

Since $\bar{R}(x) = \dfrac{R(x)}{x}$ we have

$$R(x) = (x)\bar{R}(x)$$

$$= x(250 + 45x - x^2)$$

$$= 250x + 45x^2 - x^3 \text{ (in dollars)}$$

Hence, $P(x) = R(x) - C(x)$

$$= (250x + 45x^2 - x^3) - (x^2 + 160x + 2000)$$

$$= 250x + 45x^2 - x^3 - x^2 - 160x - 2000$$

$$= -x^3 + 44x^2 + 90x - 2000 \text{ (in dollars)}$$

and

$$P'(x) = \dfrac{d}{dx}(P(x)) = \dfrac{d}{dx}(-x^3 + 44x^2 + 90x - 2000)$$

$$= -3x^2 + 88x + 90 \text{ (in dollars) } (\textbf{Marginal profit function})$$

20. $C(x) = 4x + 115$ and $P(x) = 16x - x^2 - 125$ (in dollars)

$R(x) = P(x) + C(x)$ (in dollars)

$$= (16x - x^2 - 125) + (4x + 115)$$

$$= 16x - x^2 - 125 + 4x + 115$$

$$= 20x - x^2 - 10 \text{ (in dollars)}$$

But, $R(x) = xp$

$$20x - x^2 - 10 = xp$$

$$\dfrac{20x - x^2 - 10}{x} = p$$

$$20 - x - \dfrac{10}{x} = p$$

Hence, the demand equation is given by $p = 20 - x - \dfrac{10}{x}$ where p is in dollars. (**Demand equation**)

Grade Yourself

Circle the numbers of the questions you missed, then fill in the total incorrect for each topic. If you answered more than three questions incorrectly, you need to focus on that topic. (If a topic has less than three questions and you had at least one wrong, we suggest you study that topic also. Read your textbook, a review book, or ask your teacher for help.)

Subject: Marginal Analysis

Topic	Question Numbers	Number Incorrect
Cost function	1, 2, 3	
Average cost function	4, 5	
Marginal cost function	6, 7	
Revenue function	8	
Average revenue function	9, 10	
Marginal revenue function	11, 12, 13	
Profit function	14, 15, 18	
Average profit function	16, 17	
Marginal profit function	19	
Demand equation	20	

Applications of Differentiation: Graphing

5

 Test Yourself

5.1 The First Derivative and Graphs

We know that if y = f(x) where f is a differentiable function in x, then f'(x) will give us the slope of the graph of f at the poi2ly if, given any two values x_1 and x_2 in I such that $x_1 < x_2$, then $f(x_1) < f(x_2)$. If f is increasing on I and differentiable on I, then f'(x) > 0 for all x in I. (The graph of an increasing function is "rising" from left to right.

Definition: A function is a **decreasing function** on the interval I if and only if, given any two values x_1 and x_2 in I such that $x_1 < x_2$, then $f(x_1) > f(x_2)$. If f is a decreasing function on I and differentiable on I, then f'(x) < 0 for all x in I. (The graph of a decreasing function is "falling" from left to right.)

Definition: A function is a **constant-valued function** on the interval I if and only if, given any two values x_1 and x_2 in I, then $f(x_1) = f(x_2)$. If f is a constant-valued function on I, then f'(x) = 0 for all x in I. (The graph of a constant- valued function is a horizontal line.)

In Exercises 1-3, determine whether the given function is increasing, decreasing, or neither increasing nor decreasing on its domain.

1. f(x) = 2x + 7

2. $h(x) = x^3 - 2x^2$

3. $p(x) = \dfrac{1}{x - 2}$

Note: If f is a continuous function on an interval I and f(x) ≠ 0 for all x in I, then either f(x) > 0 or f(x) < 0 for all x in I.

In Exercises 4-6, determine all intervals where the given function is increasing and all intervals where the given function is decreasing.

4. $f(x) = x^2 - 1$

5. $h(x) = x + \dfrac{2}{x}$

6. $q(x) = \dfrac{x^2}{1 - x^2}$

5.2 Extrema and the First Derivative

Definition: A function f has **a relative minimum** when x = x_1 if there is an open interval containing x_1 and such that $f(x_1) \leq f(x)$ for all x in the interval. (See Fig. 5.1a.)

Definition: A function f has **a relative maximum** when x = x_2 if there is an open interval containing x_2 and such that $f(x_2) \geq f(x)$ for all x in the interval. (See Fig. 5.1b.)

Note: If a differential function f has a relative extremum (relative minimum or relative maximum) when x = c, then either f'(c) = 0 or else f'(c) is not defined. Each relative extremum occurs at a value of x for which the sign of f'(x) changes from positive to negative or from negative to positive (even if f'(x) is not defined there).

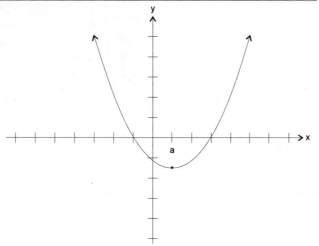

Fig. 5.1a: Relative Minimum at x = a

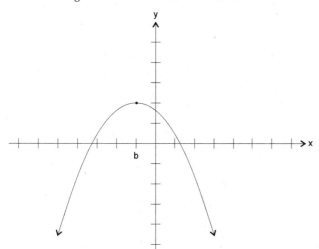

Fig. 5.1b: Relative Maximum at x = b

In Exercises 7-12, determine all relative extrema (plural of extremum) values and relative extrema points.

7. $f(x) = x^2 - 4x + 7$

8. $f(x) = 3 + x - x^2$

9. $f(x) = x^3 - 12x + 2$

10. $f(x) = \dfrac{2}{x-1}$

11. $f(x) = 2x^3 - 3x^2$

12. $f(x) = 5x - 6x^{2/3}$

In addition to relative extrema, a function may also have absolute extrema.

Definition: A function f has an **absolute minimum** at $x = x_1$ if and only if $f(x_1) \le f(x)$ for all x in the domain of f.

Definition: A function f has an **absolute maximum** at $x = x_2$ if and only if $f(x_2) \ge f(x)$ for all x in the domain of f.

Theorem: If f is continuous on [a, b], then f has both an absolute minimum value and an absolute maximum value on [a, b].

In Exercises 13-15, determine the absolute extrema for the given functions on the indicated interval.

13. $f(x) = 5 - 3x$; $[-2, 3]$

14. $f(x) = \dfrac{x^2}{x^2 + 2}$; $[-3, 0]$

15. $f(x) = x^3 - 4x^2$; $[-3, 3]$

5.3 Concavity and the Second Derivative

Definition: Let f be a differentiable function on (a, b). The graph of f is **concave upward** on (a, b) if f' is increasing on (a, b). (See Fig. 5.2a.)

Definition: Let f be a differentiable function on (a, b). The graph of f is **concave downward** on (a, b) if f' is decreasing on (a, b). (See Fig. 5.2b.)

Note: In the above definitions it is f' and not f that is increasing or decreasing on (a, b).

It follows from the above definitions that if f is a function whose second derivative exists on (a, b), then the graph of f is concave upward if f''(x) > 0 on (a, b) and the graph of f is concave downward if f''(x) < 0 on (a, b).

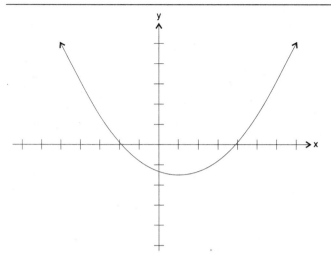

Fig. 5.2a: Concave Upward

Definition: A point on a graph where the concavity changes from concave upward to concave downward or from concave downward to concave upward is called a **point of inflection**. (See Fig. 5.3a and Fig. 5.3b.)

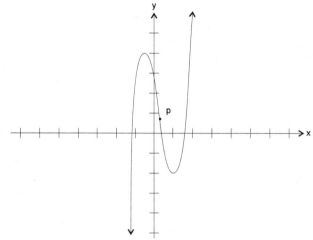

Fig. 5.3a: Point of Inflection at the Point P

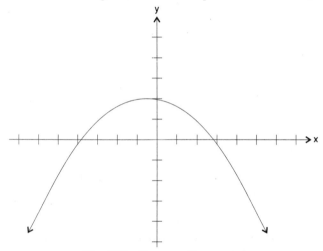

Fig. 5.2b: Concave Downward

In Exercises 16-21, determine the intervals over which the graph of the given function is concave upward and the intervals over which the graph is concave downward.

16. $f(x) = x^2 + 4$

17. $f(x) = 3x - 2x^2$

18. $f(x) = x^3 - 3x^2 + 2$

19. $f(x) = \dfrac{x^2 - 2}{2x + 3}$

20. $f(x) = x^4 - 2x^3 - 12x^2 + 12x - 9$

21. $f(x) = x^{2/3} + 8$

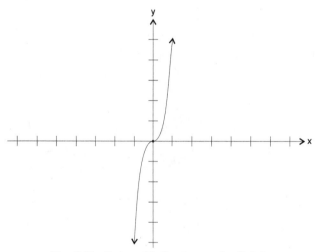

Fig. 5.3b: Point of Inflection at the Origin

In Exercises 22-24, determine all points of inflection, if any, for the graphs of the given functions.

22. $f(x) = 2x^3 - 3x^2$

23. $f(x) = x^3 - 4$

24. $f(x) = x^{2/3}(x - 2)^{1/3}$

5.4 Optimization

We often encounter problems which require determining maximum or minimum values of functions.

25. The sum of two numbers is 20. Determine the numbers if their product is to be a maximum.

26. Determine the volume of the largest box that can be made from a square piece of metal 8 inches on a side, by cutting equal squares from the corners and turning up the sides.

27. The profit function for the production and sale of x units of a product is given by $P(x) = 8x - \dfrac{x^2}{900} - 4000$ (in dollars). Determine the production and sale level that will maximize the profit.

28. Suppose that the cost function for producing a particular product is given by $C(x) = 0.06x^2 + 6x + 600$ (in dollars). Determine the level of production such that the average cost per unit is a minimum.

29. Suppose that the revenue function for the production and sale of a particular product is given by $R(x) = 5x^2 - \dfrac{x^3}{375}$ (in dollars). Determine the production and sale level in order to maximize the marginal revenue function.

5.5 Differentials

Definition: If $y = f(x)$ defines a differentiable function in x, then the **differential of y**, denoted by dy, is defined to be $dy = f'(x)\, dx$. d approximates the exact change in y when x is changed by the amount dx.

In Exercises 30-35, determine a simplified expression for dy.

30. $y = f(x) = \sqrt{3 - x}$

31. $y = f(x) = x^2 - 3x$

32. $y = f(x) = \dfrac{2x - 1}{x + 2}$

33. $y = f(x) = \sqrt{16 - x^2}$

34. $y = f(x) = (2x - 3)^2(3x - 2)^3$

35. $y = f(x) = \dfrac{1 - 2x}{x^2 + 5}$

 # Check Yourself

1. $f(x) = 2x + 7$ for all real values of x

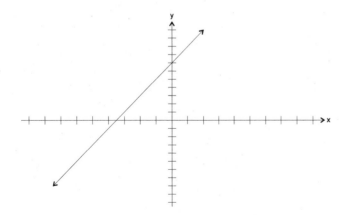

$f'(x) = \dfrac{d}{dx}(2x + 7) = 2$ for all real values of x.

Since $f'(x) = 2 > 0$ for all real values of x, then f is an increasing function on its domain. (**First derivative and graphs**)

2. $h(x) = x^3 - 2x^2$ for all real values of x

$h'(x) = \dfrac{d}{dx}(x^3 - 2x^2) = 3x^2 - 4x$ for all real values of x.

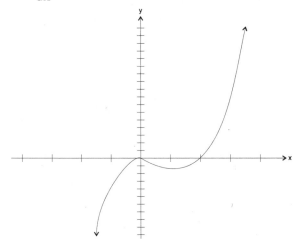

If $x = 0$, then $h'(x) = 0$. If $x = 2$, $h'(x) = 12 - 8 > 0$. If $x = 1$, $h'(x) = 3 - 4 < 0$.

Hence, h is neither increasing nor decreasing everywhere on its domain. **(First derivative and graphs)**

3. $p(x) = \dfrac{1}{x-2}$ for all real values of $x \neq 2$

$p'(x) = \dfrac{d}{dx}\left(\dfrac{1}{x-2}\right) = \dfrac{(x-2)\dfrac{d}{dx}(1) - (1)\dfrac{d}{dx}(x-2)}{(x-2)^2}$

$= \dfrac{(x-2)(0) - (1)(1)}{(x-2)^2} = \dfrac{-1}{(x-2)^2}$ for all real values of $x \neq 2$.

If $x \neq 2$, $x - 2 \neq 0$, and $(x-2)^2 > 0$.

Therefore, $\dfrac{-1}{(x-2)^2} < 0$ for all real values of $x \neq 2$.

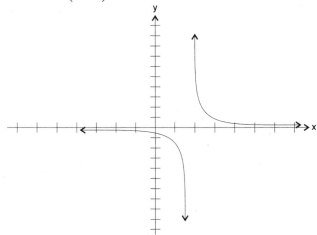

Hence, p is a decreasing function everywhere on its domain. **(First derivative and graphs)**

4. $f(x) = x^2 - 1$ for all real values of x

$$f'(x) = \frac{d}{dx}(x^2 - 1) = 2x \text{ for all real values of x.}$$

Set $f'(x) = 0$: $2x = 0$ or $x = 0$ (a boundary number). Create a sign chart for $f'(x)$ using the boundary number 0.

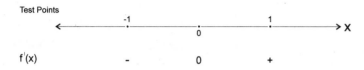

Using the note following Exercise 3, the sign of $f'(x)$ will be the same for all x in $(-\infty, 0)$, and the same for all x in $(0, +\infty)$. Take test points -1 and 1, respectively, in these intervals.

$f'(-1) = 2(-1) < 0$. Hence, $f'(x) < 0$ for all x in $(-\infty, 0)$.

$f'(1) = 2(1) > 0$. Hence, $f'(x) > 0$ for all x in $(0, +\infty)$.

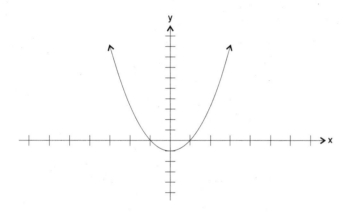

Therefore, f is increasing on $(0, +oo)$ and is decreasing on $(-\infty, 0)$. **(First derivative and graphs)**

5. $h(x) = x + \frac{2}{x}$ for all real values of $x \neq 0$

$$h'(x) = \frac{d}{dx}\left(x + \frac{2}{x}\right) = 1 - \frac{2}{x^2} \text{ for all real values of } x \neq 0.$$

Set $h'(x) = 0$: $1 - \frac{2}{x^2} = 0$ or $x^2 - 2 = 0$ or $x = \pm\sqrt{2}$.

The boundary numbers are $\pm\sqrt{2}$. (Note that 0 is also a boundary number, since the function is not defined when $x = 0$.) Create a sign chart for $h'(x)$ using the boundary numbers $-\sqrt{2}$, 0, and $\sqrt{2}$.

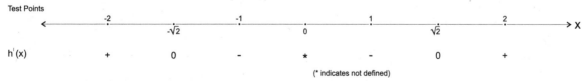

Take test points -2, -1, 1, and 2.

$h'(-2) = 1 - (1/2) > 0$. Hence, $h'(x) > 0$ for all x in $(-\infty, -\sqrt{2})$.

$h'(-1) = 1 - 2 < 0$. Hence, $h'(x) < 0$ for all x in $(-\sqrt{2}, 0)$.

$h'(1) = 1 - 2 < 0$. Hence, $h'(x) < 0$ for all x in $(0, \sqrt{2})$.

$h'(2) = 1 - (1/2) > 0$. Hence, $h'(x) > 0$ for all x in $(\sqrt{2}, +\infty)$.

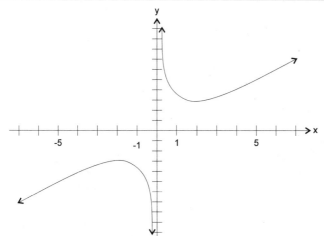

Therefore, h is increasing on $(-\infty, -\sqrt{2}) \cup (\sqrt{2}, +\infty)$ and is decreasing on $(-\sqrt{2}, 0) \cup (0, \sqrt{2})$. **(First derivative and graphs)**

6. $q(x) = \dfrac{x^2}{1 - x^2}$ for all real values of $x \neq \pm 1$

$$q'(x) = \frac{d}{dx}\left(\frac{x^2}{1 - x^2}\right) = \frac{(1 - x^2)\dfrac{d}{dx}(x^2) - x^2\dfrac{d}{dx}(1 - x^2)}{(1 - x^2)^2}$$

$$= \frac{(1 - x^2)(2x) - x^2(0 - 2x)}{(1 - x^2)^2} = \frac{2x - 2x^3 + 2x^3}{(1 - x^2)^2}$$

$$= \frac{2x}{(1 - x^2)^2} \text{ for all real values of } x \neq \pm 1.$$

Set $q'(x) = 0$: $\dfrac{2x}{(1 - x^2)^2} = 0$; $2x = 0$ or $x = 0$. The boundary numbers are $-1, 0$, and 1. Create a sign chart for $q'(x)$ using these boundary numbers.

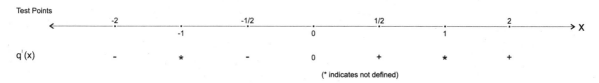

Take test points $-2, -1/2, 1/2$, and 2.

$q'(-2) = (-4)/9 < 0$. Hence, $q'(x) < 0$ for all x in $(-\infty, -1)$.

$q'(-1/2) = (-1)/[9/16] < 0$. Hence, $q'(x) < 0$ for all x in $(-1, 0)$.

$q'(1/2) = 1/[9/16] > 0$. Hence, $q'(x) > 0$ for all x in $(0, 1)$.

$q'(2) = 4/9 > 0$. Hence, $q'(x) > 0$ for all x in $(1, +\infty)$.

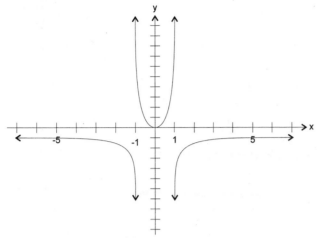

Therefore, q is increasing on (0, 1) and on (1, +∞) and is decreasing on (−∞, −1) and on (−1, 0). (**First derivative and graphs**)

7. $f(x) = x^2 - 4x + 7$ for all real values of x

$f'(x) = \dfrac{d}{dx}(x^2 - 4x + 7) = 2x - 4$ for all real values of x.

Set $f'(x) = 0$: $2x - 4 = 0$ or $x = 2$ (a boundary number). Create a sign chart for $f'(x)$ using this boundary number.

Take test points 0 and 3.

$f'(0) = (2)(0) - 4 < 0$. [f is decreasing on (−∞, 2).]

$f'(3) = (2)(3) - 4 > 0$. [f is increasing on (2, +∞).]

Since $f'(x)$ changes sign from negative to positive at $x = 2$, a relative minimum occurs at $x = 2$.

$f(2) = (2)^2 - 4(2) + 7 = 4 - 8 + 7 = 3$ (a relative minimum *value*).

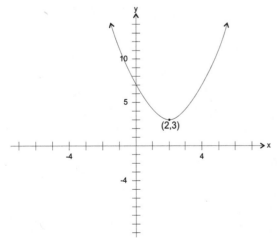

The relative minimum *value* is 3 and occurs at the relative minimum *point* (2, 3). (**Extrema and the first derivative**)

8. $f(x) = 3 + x - x^2$ for all real values of x

$f'(x) = \dfrac{d}{dx}(3 + x - x^2) = 1 - 2x$ for all real values of x.

Set $f'(x) = 0$: $1 - 2x = 0$ or $x = 1/2$ (a boundary number). Create a sign chart for $f'(x)$ using this boundary number.

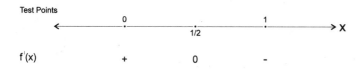

Take test points 0 and 1.

$f'(0) = 1 - 2(0) > 0$. [f is increasing on $(-\infty, 1/2)$].

$f'(1) = 1 - 2(1) < 0$. [f is decreasing on $(1/2, +\infty)$].

Since $f'(x)$ changes sign from positive to negative at $x = 1/2$, a relative maximum occurs at $x = 1/2$.

$f(1/2) = 3 + (1/2) - (1/4) = 13/4$ (a relative maximum *value*).

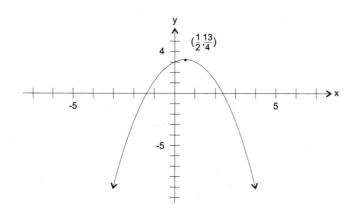

The relative maximum *value* is 13/4 and occurs at the relative maximum *point* (1/2, 13/4). (**Extrema and the first derivative**)

9. $f(x) = x^3 - 12x + 2$ for all real values of x

$f'(x) = \dfrac{d}{dx}(x^3 - 12x + 2) = 3x^2 - 12$ for all real values of x.

Set $f'(x) = 0$: $3x^2 - 12 = 0$ or $x^2 = 4$ or $x = \pm 2$ (boundary numbers). Create a sign chart for $f'(x)$ using these boundary numbers.

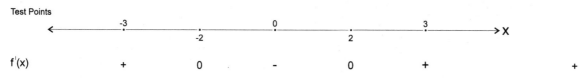

Take test points −3, 0, and 3.

$f'(-3) = 27 - 12 > 0$. [f is increasing on $(-\infty, -2)$.]

$f'(0) = 0 - 12 < 0$. [f is decreasing on $(-2, 2)$.]

$f'(3) = 27 - 12 > 0$. [f is increasing on $(2, +\infty)$.]

Since f'(x) changes sign from positive to negative at x = −2, a relative maximum occurs at x = −2.

f(−2) = (−2)³ − 12(−2) + 2 = 18 (a relative maximum *value*).

Since f'(x) changes sign from negative to positive at x = 2, a relative minimum occurs at x = 2.

f(2) = (2)³ − 12(2) + 2 = −14 (a relative minimum *value*).

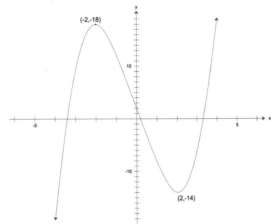

The relative maximum *value* is 18 and occurs at the relative maximum *point* (−2, 18). The relative minimum *value* is −14 and occurs at the relative minimum *point* (2, −14). **(Extrema and the first derivative)**

10. $f(x) = \dfrac{2}{x-1}$ for all real values of x ≠ 1

$$f'(x) = \frac{d}{dx}\left(\frac{2}{x-1}\right) = \frac{d}{dx}[2(x-1)^{-1}] = -2(x-2)^{-2} = \frac{-2}{(x-1)^2} \text{ all real values of } x \neq 1.$$

Set $f'(x) = 0$: $\dfrac{-2}{(x-1)^2} = 0$; no solution.

There are no real values of x for which f'(x) = 0. However, f'(x) is not defined at x = 1 (a boundary number). Create a sign chart for f'(x) using this boundary number.

Take test points 0 and 2.

f'(0) = −2 < 0. [f is decreasing on (−∞, 1).]

f'(2) = −2 < 0. [f is decreasing on (1, +∞).]

There are no changes of sign for f(x). Hence, there are no relative extrema values or relative extrema points. **(Extrema and the first derivative)**

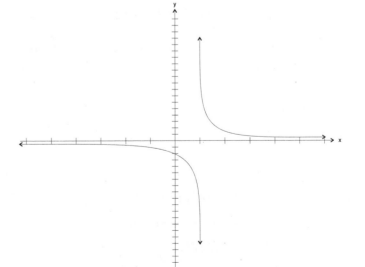

11. $f(x) = 2x^3 - 3x^2$ for all real values of x

$f'(x) = \dfrac{d}{dx}(2x^3 - 3x^2) = 6x^2 - 6x$ for all real values of x.

Set $f'(x) = 0$: $6x^2 - 6x = 0$ or $6x(x - 1) = 0$ or $x = 0, 1$ (boundary numbers). Create a sign chart for $f'(x)$ using these boundary numbers.

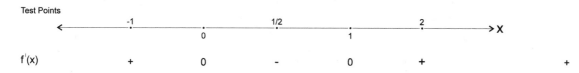

Test Points

f'(x)   +   0   -   0   +   +

Take test points −1, 1/2, and 2.

$f'(-1) = 6 + 6 > 0$. [f is increasing on $(-\infty, 0)$.]

$f'(1/2) = (3/2) - 3 < 0$. [f is decreasing on $(0, 1)$.]

$f'(2) = 24 - 12 > 0$. [f is increasing on $(1, +\infty)$.]

Since $f'(x)$ changes sign from positive to negative at $x = 0$, a relative maximum occurs at $x = 0$.

$f(0) = 2(0)^3 - 3(0)^2 = 0$ (a relative maximum *value*).

Since $f'(x)$ changes sign from negative to positive at $x = 1$, a relative minimum occurs at $x = 1$.

$f(1) = 2(1)^3 - 3(1)^2 = -1$ (a relative minimum *value*).

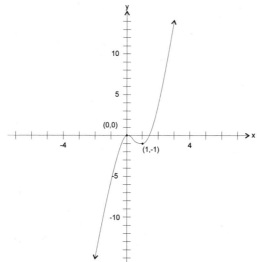

The relative maximum *value* is 0 and occurs at the relative maximum *point* $(0, 0)$. The relative minimum *value* is −1 and occurs at the relative minimum *point* $(1, -1)$. (**Extrema and the first derivative**)

12. $f(x) = 5x - 6x^{2/3}$ for all x

$f'(x) = \dfrac{d}{dx}(5x - 6x^{2/3}) = 5 - \dfrac{4}{x^{1/3}}$ for all real values of $x \neq 0$.

Set $f'(x) = 0$: $5 - \dfrac{4}{x^{1/3}} = 0$ or $5x^{1/3} = 4$ or $x = \dfrac{64}{125}$ (The boundary numbers are 0 and $\dfrac{64}{125}$).

Create a sign chart for $f'(x)$ using these boundary numbers.

Test Points

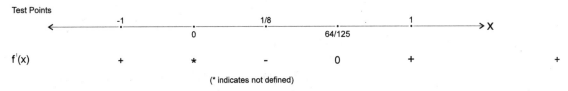

| f'(x) | + | * | – | 0 | + | + |

(* indicates not defined)

Take test points –1, 1/8, and 1.

f'(–1) = 5 – 4 > 0. [f is increasing on (–∞, 0).]

f'(1/8) = 5 – 8 < 0. [f is decreasing on (0, 64/125).]

f'(1) = 5 – 4 > 0. [f is increasing on (64/125, +∞).]

Since f'(x) changes sign from positive to negative at x = 0, a relative maximum occurs at x = 0.

f(0) = 5(0) – 6(0)2/3 = 0 (a relative maximum *value*).

Since f'(x) changes sign from negative to positive at x = 64/125, a relative minimum occurs at x = 64/125.

f(64/125) = 5(64/125) – 6(64/125)$^{2/3}$ = –32/25 (a relative minimum *value*).

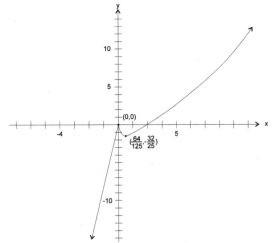

The relative maximum *value* is 0 and occurs at the relative maximum *point* (0, 0). The relative minimum *value* is –32/25 and occurs at the relative minimum *point* $\left(\dfrac{64}{125}, \dfrac{-32}{25} \right)$. **(Extrema and the first derivative)**

13. f(x) = 5 – 3x; [–2, 3]

$f'(x) = \dfrac{d}{dx}(5 - 3x) = -3$ for all x in [–2, 3].

Set f'(x) = 0: Since –3 ≠ 0, there are no solutions for the equation. Hence, there are no relative extrema. Test the function values at the endpoints of the interval [–2, 3].

f(–2) = 5 – 3(–2) = 11 (absolute maximum *value*)

f(3) = 5 – 3(3) = –4 (absolute minimum *value*)

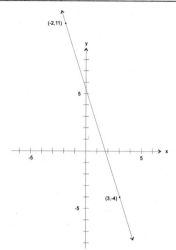

Hence, the absolute maximum value is 11 and occurs at the point (−2, 11). The absolute minimum value is −4 and occurs at the point (3, −4). (**Extrema and the first derivative**)

14. $f(x) = \dfrac{x^2}{x^2 + 2}$; [−3, 0]

$$f'(x) = \frac{d}{dx}\left(\frac{x^2}{x^2+2}\right) = \frac{(x^2+2)\frac{d}{dx}(x^2) - x^2\frac{d}{dx}(x^2+2)}{(x^2+2)^2}$$

$$= \frac{(x^2+2)(2x) - x^2(2x)}{(x^2+2)^2} = \frac{2x^3 + 4x - 2x^3}{(x^2+2)^2} = \frac{4x}{(x^2+2)^2} \text{ for all x in } [-3, 0].$$

Set f′(x) = 0: $\dfrac{4x}{(x^2+2)^2} = 0$ or x = 0 (a boundary number). Create a sign chart for f(x) using this boundary number. Since 0 is the right endpoint of the interval [−3, 0], there are no relative extrema. Test the function values at the endpoints of the interval [−3, 0].

f(−3) = 9/11. (the absolute maximum *value*)

f(0) = 0/2 = 0. (the absolute minimum *value*)

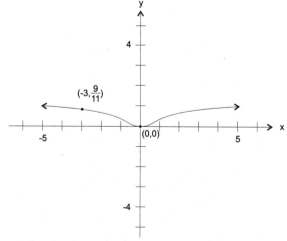

Hence, the absolute maximum *value* of the function on the given interval is 9/11 and occurs when x = −3. The absolute minimum *value* of the function on the given interval is 0 and occurs when x = 0. (**Extrema and the first derivative**)

15. $f(x) = x^3 - 4x^2$; [−3, 3]

$f'(x) = \dfrac{d}{dx}(x^3 - 4x^2) = 3x^2 - 8x$ for all real values of x in [−3, 3].

Set $f'(x) = 0$: $3x^2 - 8x = 0$ or $x(3x - 8) = 0$ or x = 0, 8/3 (the boundary numbers). Create a sign chart for $f'(x)$ using these boundary numbers.

Take test points −1, 1, and 2.8 .

$f'(-1) = 3 + 8 > 0$. [f is increasing on (−3, 0).]

$f'(1) = 3 − 8 < 0$. [f is decreasing on (0, 8/3).]

$f'(2.8) = 23.52 − 22.4 > 0$. [f is increasing on (8/3, 3).]

Since $f'(x)$ changes sign from positive to negative at x = 0, a relative maximum occurs at x = 0. Since $f'(x)$ changes sign from negative to positive at x = 8/3, a relative minimum occurs at x = 8/3. Test the function values at the endpoints of the interval [−3, 3] and also at the values where relative extrema occur.

$f(-3) = -27 - 36 = -63$ (absolute minimum *value*)

$f(0) = 0 - 0 = 0$ (absolute maximum *value*)

$f(8/3) = (512/27) - (256/9) = -256/27$

$f(3) = 27 - 36 = -9$

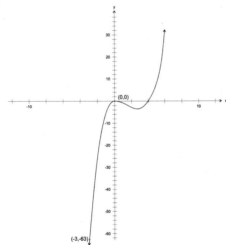

Hence, the absolute maximum *value* of the function on the given interval is 0 and occurs when x = 0. The absolute minimum *value* of the function on the given interval is −63 and occurs when x = −3. (**Extrema and the first derivative**)

16. $f(x) = x^2 + 4$ for all real values of x.

$f'(x) = \dfrac{d}{dx}(x^2 + 4) = 2x$ for all real values of x.

$f''(x) = \dfrac{d}{dx}(2x) = 2$ for all real values of x.

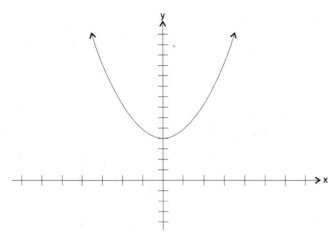

Since $f''(x) = 2 > 0$ for all real values of x, then the graph of f is concave upward everywhere on its domain. **(Concavity and the second derivative)**

17. $f(x) = 3x - 2x^2$ for all real values of x.

$f'(x) = \dfrac{d}{dx}(3x - 2x^2) = 3 - 4x$ for all real values of x.

$f''(x) = \dfrac{d}{dx}(3 - 4x) = -4$ for all real values of x.

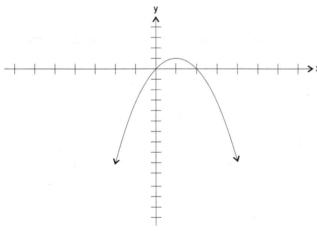

Since $f''(x) = -4 < 0$ for all real values of x, then the graph of f is concave downward everywhere on its domain. **(Concavity and the second derivative)**

18. $f(x) = x^3 - 3x^2 + 2$ for all real values of x.

$f'(x) = \dfrac{d}{dx}(x^3 - 3x^2 + 2) = 3x^2 - 6x$ for all real values of x.

$f''(x) = \dfrac{d}{dx}(3x^2 - 6x) = 6x - 6$ for all real values of x.

Set $f''(x) = 0$: $6x - 6 = 0$ or $x = 1$ (a boundary number). Create a sign chart for $f''(x)$ using this boundary number.

Test Points

f″(x)   −   0   +

Take test points 0 and 2.

f″(0) = 6(0) − 6 < 0; the graph of f is concave downward on (−∞, 1).

f″(2) = 6(2) − 6 > 0; the graph of f is concave upward on (1, +∞).

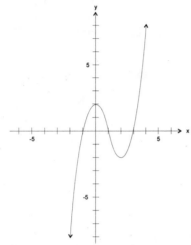

Hence, the graph of f is concave downward on (−∞, 1) and is concave upward on (1, +∞). **(Concavity and the second derivative)**

19. $f(x) = \dfrac{x^2 - 2}{2x + 3}$ for all real values of x ≠ −3/2.

$$f'(x) = \frac{d}{dx}\left(\frac{x^2 - 2}{2x + 3}\right) = \frac{(2x + 3)\frac{d}{dx}(x^2 - 2) - (x^2 - 2)\frac{d}{dx}(2x + 3)}{(2x + 3)^2}$$

$$= \frac{(2x + 3)(2x) - (x^2 - 2)(2)}{(2x + 3)^2}$$

$$= \frac{4x^2 + 6x - 2x^2 + 4}{(2x + 3)^2}$$

$$= \frac{2x^2 + 6x + 4}{(2x + 3)^2} \text{ for all } x \neq -3/2.$$

$$f''(x) = \frac{d}{dx}\left(\frac{2x^2 + 6x + 4}{(2x + 3)^2}\right)$$

$$= \frac{(2x + 3)^2\frac{d}{dx}(2x^2 + 6x + 4) - (2x^2 + 6x + 4)\frac{d}{dx}(2x + 3)^2}{(2x + 3)^4}$$

$$= \frac{(2x + 3)^2(4x + 6) - (2x^2 + 6x + 4)(2)(2x + 3)(2)}{(2x + 3)^4}$$

$$= \frac{(4x+6)[(2x+3)^2 - 2(2x^2+6x+4)]}{(2x+3)^4}$$

$$= \frac{(4x+6)(4x^2+12x+9) - 4x^2 - 12x - 8)}{(2x+3)^4}$$

$$= \frac{(4x+6)(1)}{(2x+3)^4} = \frac{2(2x+3)}{(2x+3)^4} = \frac{2}{(2x+3)^3} \text{ for all real values of } x \neq -3/2.$$

Set $f''(x) = 0$: Since $\dfrac{2}{(2x+3)^3} \neq 0$, there are no solutions.

Hence, the only boundary number is $x = -3/2$. Create a sign chart for $f''(x)$ using the boundary number $x = -3/2$.

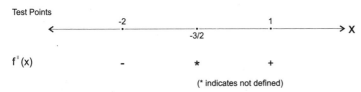

Take test points -2 and 1.

$f''(-2) = 2/(-1) = -2 < 0$; the graph of f is concave downward on $(-\infty, -3/2)$.

$f''(1) = 2/125 > 0$; the graph of f is concave upward on $(-3/2, +\infty)$.

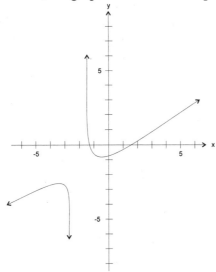

Hence, the graph of f is concave downward on $(-\infty, -3/2)$ and is concave upward on $(-3/2, +\infty)$.
(Concavity and the second derivative)

20. $f(x) = x^4 - 2x^3 - 12x^2 + 12x - 9$ for all real values of x.

$$f'(x) = \frac{d}{dx}(x^4 - 2x^3 - 12x^2 + 12x - 9)$$

$$= 4x^3 - 6x^2 - 24x + 12 \text{ for all real values of } x.$$

$$f''(x) = \frac{d}{dx}(4x^3 - 6x^2 - 24x + 12) = 12x^2 - 12x - 24 \text{ for all real values of } x.$$

Set $f''(x) = 0$: $12x^2 - 12x - 24 = 0$ or $x^2 - x - 2 = 0$ or $(x-2)(x+1) = 0$ or $x = -1, 2$ (boundary numbers). Create a sign chart for $f''(x)$ using these boundary numbers.

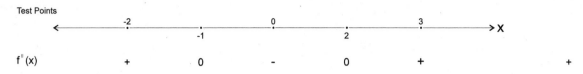

Test Points

f''(x)   +   0   -   0   +   +

Take test points −2, 0, and 3.

f''(−2) = 48 + 24 − 24 > 0; the graph of f is concave upward on (−∞, −1).

f''(0) = 0 − 0 − 24 < 0; the graph of f is concave downward on (−1, 2).

f''(3) = 108 − 36 − 24 > 0; the graph of f is concave upward on (2, +∞).

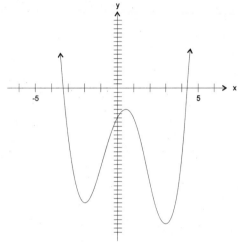

Hence, the graph of f is concave upward on (−∞, −1) ∪ (2, +∞) and is concave downward on (−1, 2). (**Concavity and the second derivative**)

21. f(x) = x$^{2/3}$ + 8 for all real values of x.

$$f'(x) = \frac{d}{dx}(x^{2/3} + 8) = (2/3)x^{-1/3}$$ for all real values of x ≠ 0.

$$f''(x) = \frac{d}{dx}[(2/3)x^{-1/3}].$$

$$= (2/3)(-1/3)x^{-4/3} = (-2/9)x^{-4/3}$$ for all x ≠ 0.

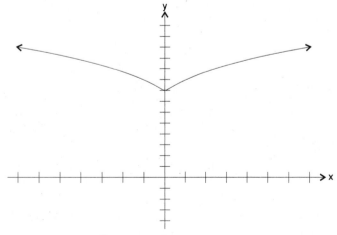

Since f''(x) = $\dfrac{-2}{9x^{4/3}}$ < 0 for all x ≠ 0, the graph of f is concave downward everywhere on its domain.

(**Extrema and the second derivative**)

22. $f(x) = 2x^3 - 3x^2$ for all real values of x.

$f'(x) = \dfrac{d}{dx}(2x^3 - 3x^2) = 6x^2 - 6x$ for all real values of x.

$f''(x) = \dfrac{d}{dx}(6x^2 - 6x) = 12x - 6$ for all real values of x.

Set $f''(x) = 0$: $12x - 6 = 0$ or $x = 1/2$ (a boundary number).

Create a sign chart for $f''(x)$ using this boundary number.

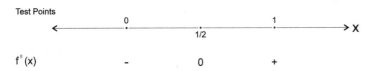

Take test points 0 and 1.

$f''(0) = 0 - 6 < 0$; the graph of f is concave downward on $(-\infty, 1/2)$.

$f''(1) = 12 - 6 > 0$; the graph of f is concave upward on $(1/2, +\infty)$.

Since $f''(x)$ changes sign at $x = 1/2$, the graph of f changes concavity at $x = 1/2$. Hence, there is a point of inflection at $x = 1/2$.

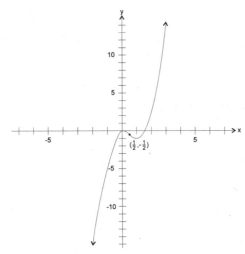

$f(1/2) = 2(1/8) - 3(1/4) = (1/4) - (3/4) = -1/2$.

The point of inflection is at $(1/2, -1/2)$. **(Concavity and the second derivative)**

23. $f(x) = x^3 - 4$ for all real values of x.

$f'(x) = \dfrac{d}{dx}(x^3 - 4) = 3x^2$ for all real values of x.

$f''(x) = \dfrac{d}{dx}(3x^2) = 6x$ for all real values of x.

Set $f''(x) = 0$: $6x = 0$ or $x = 0$ (a boundary number).

Create a sign chart for $f''(x)$ using this boundary number.

Take test points -1 and 1.

$f''(-1) = -6 < 0$; the graph of f is concave downward on $(-\infty, 0)$.

$f''(1) = 6 > 0$; the graph of f is concave upward on $(0, +\infty)$.

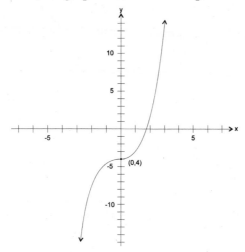

Since $f''(x)$ changes sign at $x = 0$, the graph of f changes concavity at $x = 0$. Hence, there is a point of inflection at $x = 0$.

$f(0) = (0)^3 - 4 = -4$.

The point of inflection is at $(0, -4)$. (**Concavity and the second derivative**)

24. $f(x) = x^{2/3}(x - 2)^{1/3}$ for all real values of x.

$$f'(x) = \frac{d}{dx}[x^{2/3}(x - 2)^{1/3}]$$

$$= \frac{d}{dx}[x^{2/3}(x - 2)^{1/3}]$$

$$= \frac{d}{dx}[x^2(x - 2)]^{1/3} = \frac{d}{dx}[(x^2 - 2x^2)^{1/3}]$$

$$= (1/3)(x^3 - 2x^2)^{-2/3}(3x^2 - 4x) \text{ for all real values of } x \neq 0, 2.$$

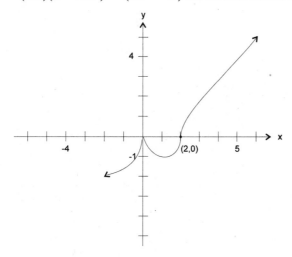

$$f''(x) = \frac{d}{dx}[(1/3)(x^3 - 2x^2)^{-2/3}(3x^2 - 4x)]$$

$$= \frac{1}{3}(x^3 - 2x^2)^{-2/3}\frac{d}{dx}(3x^2 - 4x) + (3x^2 - 4x)\frac{d}{dx}[\frac{1}{3}(x^3 - 2x^2)^{-2/3}]$$

$$= \frac{1}{3}(x^3 - 2x^2)^{-2/3}(6x - 4) + (3x^2 - 4x)\left[\left(\frac{1}{3}\right)\left(\frac{-2}{3}\right)(x^3 - 2x^2)^{-5/3}(3x^2 - 4x)\right]$$

$$= \frac{2}{3}(3x - 2)(x^3 - 2x^2)^{-2/3} - \frac{2}{9}(3x^2 - 4x)^2(x^3 - 2x^2)^{-5/3}$$

$$= \frac{2}{3}(3x - 2)[x^2(x - 2)]^{-2/3} - \frac{2}{9}(3x^2 - 4x)^2[x^2(x - 2)]^{-5/3}$$

$$= \frac{2}{3}(3x - 2)x^{-4/3}(x - 2)^{-2/3} - \frac{2}{9}x^2(3x - 4)^2\,x^{-10/3}(x - 2)^{-5/3}$$

$$= \frac{2}{3}(3x - 2)x^{-4/3}(x - 2)^{-2/3} - \frac{2}{9}x^{-4/3}(3x - 4)^2(x - 2)^{-5/3}$$

$$= \frac{2}{9}x^{-4/3}(x - 2)^{-5/3}[3(3x - 2)(x - 2) - (3x - 4)^2]$$

$$= \frac{2}{9}x^{-4/3}(x - 2)^{-5/3}[9x^2 - 24x + 12 - 9x^2 + 24x - 16]$$

$$= \frac{2}{9}x^{-4/3}(x - 2)^{-5/3}(-4)$$

$$= \frac{-8}{9}x^{-4/3}(x - 2)^{-5/3} \text{ for any real values of } x \neq 0, 2.$$

Set f''(x) = 0: Since f''(x) ≠ 0 for all real values of x, the boundary numbers are only 0 and 2. Create a sign chart for f''(x) using these boundary numbers.

(* indicates not defined)

Take test points −1, 1, and 3.

f''(−1) = (−8/9)(1)[1/(−3 $\sqrt[3]{9}$)] > 0; the graph of f is concave upward on (−∞, 0).

f''(1) = (−8/9)(1)(−1) > 0; the graph of f is concave upward on (0, 2).

f''(3) = (−8/9)[1/(3 $\sqrt[3]{3}$)](1) < 0; the graph of f is concave downward on (2, +∞).

Since f''(x) changes sign at x = 2, the graph of f changes concavity at x = 2. Hence, there is a point of inflection at x = 2.

f(2) = (2)$^{2/3}$(0) = 0

The point of inflection is at (2, 0). **(Concavity and the second derivative)**

25. Let n = one number and 20 − n = the other number.

We want to maximize n(20 − n).

Let P(n) = n(20 − n) = 20n − n^2.

Determine $P'(n) = \dfrac{d}{dn}(20n - n^2) = 20 - 2n$.

Set $P'(n) = 0$: $20 - 2n = 0$ or $n = 10$.

Determine $P''(n)$: $P''(n) = \dfrac{d}{dn}(20 - 2n) = -2$ for all n.

Since $P''(10) = -2 < 0$, P(n) will be maximized at $n = 10$.

Since $n = 10$, then $20 - n = 20 - 10 = 10$.

The required numbers are 10 and 10 and the maximum product is $(10)(10) = 100$. **(Optimization)**

26. Let x be a side of a corner square (in inches). Then,

$$V(x) = x(8 - 2x)(8 - 2x) \ (\text{in.}^3)$$
$$= x(64 - 32x + 4x^2)$$
$$= 64x - 32x^2 + 4x^3$$

Determine $V'(x)$: $V'(x) = \dfrac{d}{dx}(64x - 32x^2 + 4x^3)$

$$= 64 - 64x + 12x^2.$$

Set $V'(x) = 0$: $64 - 64x + 12x^2 = 0$

$$3x^2 - 16x + 16 = 0$$
$$(3x - 4)(x - 4) = 0$$
$$3x - 4 = 0 \text{ or } x - 4 = 0$$
$$x = 4/3 \qquad x = 4$$

We seem to have two solutions. However, from the figure above, we note that x cannot be equal to 4. (Why?) Hence, the solution seems to be $x = 4/3$.

Determine $V''(x)$: $V''(x) = \dfrac{d}{dx}(64 - 64x + 12x^2)$

$$= -64 + 24x \text{ for all real values of x.}$$

$V''(4/3) = -64 + 24(4/3) = -32 < 0$. Since the second derivative of V(x) is negative when $x = 4/3$, the graph of the function will be concave down. Therefore, $x = 4/3$ will produce the maximum volume.

$$V(4/3) = (4/3)[8 - 2(4/3)]^2 = (4/3)[8 - (8/3)]^2$$
$$= 1024/27 \text{ in.}^3$$

The maximum volume, $1024/27$ in.^{3}, will occur by cutting out square corners measuring 4/3 in. on a side. **(Optimization)**

27. $P(x) = 8x - \dfrac{x^2}{900} - 4000$ for all $x > 0$.

Determine $P'(x)$: $P'(x) = \dfrac{d}{dx}\left(8x - \dfrac{x^2}{900} - 4000\right) = 8 - \dfrac{x}{450}$.

Set $P'(x) = 0$: $8 - \dfrac{x}{450} = 0$ or $x = 3600$.

Determine $P''(x)$: $P''(x) = \dfrac{d}{dx}(8 - \dfrac{x}{450}) = \dfrac{-1}{450}$ for all $x > 0$.

Since $P''(3600) = \dfrac{-1}{450} < 0$, profit will be maximized when $x = 3600$ units. (The maximum profit will be $10,400.) **(Optimization)**

28. $C(x) = 0.06x^2 + 6x + 600$ (in dollars)

$\overline{C}(x) = \dfrac{C(x)}{x} = \dfrac{0.06x^2 + 6x + 600}{x} = 0.06x + 6 + \dfrac{600}{x}$ (in dollars)

Determine $\overline{C}'(x)$: $\overline{C}'(x) = \dfrac{d}{dx}(0.06x + 6 + \dfrac{600}{x})$

$= 0.06 - \dfrac{600}{x^2}$

Set $\overline{C}'(x) = 0$: $0.06 - \dfrac{600}{x^2} = 0$ or $0.06x^2 = 600$ or $x^2 = 10,000$ or $x = \pm100$. (Reject $x = -100$.)

Determine $\overline{C}''(x)$: $\overline{C}''(x) = \dfrac{d}{dx}(0.06 - \dfrac{600}{x^2}) = \dfrac{1200}{x^3}$ for all $x > 0$.

Since $\overline{C}''(100) = \dfrac{1200}{x^3} > 0$, the average cost function will be minimized when $x = 100$.

$(\overline{C}(100) = 18$ (dollars)) **(Optimization)**

29. $R(x) = 5x^2 - \dfrac{x^3}{375}$ (in dollars)

Marginal revenue $= R'(x) = \dfrac{d}{dx}(5x^2 - \dfrac{x^3}{375}) = 10x - \dfrac{x^2}{125}$.

Determine $\dfrac{d}{dx}(R'(x)) = \dfrac{d}{dx}(10x - \dfrac{x^2}{125}) = 10 - \dfrac{2x}{125}$.

Set $\dfrac{d}{dx}(R'(x)) = 0$: $10 - \dfrac{2x}{125} = 0$ or $2x = 1250$ or $x = 625$.

Determine $\dfrac{d^2}{dx^2}(R'(x)) = \dfrac{d}{dx}(10 - \dfrac{2x}{125}) = \dfrac{-2}{125}$ for all $x > 0$.

Since $\dfrac{d^2}{dx^2}(R'(625)) = \dfrac{-2}{125}$, $R'(x)$ will be maximized when $x = 625$ units. ($R'(625) = \$3125$.)
(Optimization)

30. $y = f(x) = \sqrt{3 - x}$ for all real values of $x \le 3$.

$dy = f'(x)\, dx$

$= \dfrac{d}{dx}(\sqrt{3 - x})dx$

$= \dfrac{1}{2}(3 - x)^{-1/2}(-1)\, dx$

$$= \frac{-1}{2\sqrt{3-x}} \, dx \text{ for all real values of x} < 3. \textbf{ (Differentials)}$$

31. $y = f(x) = x^2 - 3x$ for all real values of x.

$$dy = f'(x) \, dx$$

$$= \frac{d}{dx}(x^2 - 3x) \, dx$$

$$= (2x - 3) \, dx \text{ for all real values of x.} \textbf{ (Differentials)}$$

32. $y = f(x) = \dfrac{2x-1}{x+2}$ for all real values of $x \neq -2$.

$$dy = f'(x) \, dx$$

$$= \frac{d}{dx}\left(\frac{2x-1}{x+2}\right) dx$$

$$= \frac{(x+2)\dfrac{d}{dx}(2x-1) - (2x-1)\dfrac{d}{dx}(x+2)}{(x+2)^2} \, dx$$

$$= \frac{(x+2)(2) - (2x-1)(1)}{(x+2)^2} \, dx$$

$$= \frac{2x+4-2x+1}{(x+2)^2} \, dx$$

$$= \frac{5}{(x+2)^2} \, dx \text{ for all real values of } x \neq -2. \textbf{ (Differentials)}$$

33. $y = f(x) = \sqrt{16 - x^2}$ for all real values of x such that $|x| \leq 4$.

$$dy = f'(x) \, dx$$

$$= \frac{d}{dx}(\sqrt{16 - x^2}) \, dx$$

$$= \frac{1}{2}(16 - x^2)^{-1/2} \, (-2x) \, dx$$

$$= \frac{-x}{\sqrt{16 - x^2}} \, dx \text{ for all real values of x such that } |x| < 4. \textbf{ (Differentials)}$$

34. $y = f(x) = (2x - 3)^2(3x - 2)^3$ for all real values of x.

$$dy = f'(x) \, dx$$

$$= \frac{d}{dx}[(2x-3)^2(3x-2)^3] \, dx$$

$$= [(2x-3)^2\frac{d}{dx}(3x-2)^3 + (3x-2)^3\frac{d}{dx}(2x-3)^2] \, dx$$

$$= [(2x-3)^2(3)(3x-2)^2(3) + (3x-2)^3(2)(2x-3)(2)] \, dx$$

$$= (2x-3)(3x-2)^2[9(2x-3) + 4(3x-2)] \, dx$$

$$= (2x-3)(3x-2)^2(18x-27 + 12x-8) \, dx$$

$$= 5(2x-3)(6x-7)(3x-2)^2 \, dx \text{ for all real values of x.} \textbf{ (Differentials)}$$

35. $y = f(x) = \dfrac{1-2x}{x^2+5}$ for all real values of x.

$dy = f'(x)\, dx$

$\quad = \dfrac{d}{dx}\left(\dfrac{1-2x}{x^2+5}\right)dx$

$\quad = \dfrac{(x^2+5)\dfrac{d}{dx}(1-2x) - (1-2x)\dfrac{d}{dx}(x^2+5)}{(x^2+5)^2}\, dx$

$\quad = \dfrac{(x^2+5)(-2) - (1-2x)(2x)}{(x^2+5)^2}\, dx$

$\quad = \dfrac{-2x^2 - 10 - 2x + 4x^2}{(x^2+5)^2}\, dx$

$\quad = \dfrac{2x^2 - 2x - 10}{(x^2+5)^2}\, dx$ for all real values of x. **(Differentials)**

Grade Yourself

Circle the numbers of the questions you missed, then fill in the total incorrect for each topic. If you answered more than three questions incorrectly, you need to focus on that topic. (If a topic has less than three questions and you had at least one wrong, we suggest you study that topic also. Read your textbook, a review book, or ask your teacher for help.)

Subject: Applications of Differentiation; Graphing

Topic	Question Numbers	Number Incorrect
First deriative and graphs	1, 2, 3, 4, 5, 6	
Extrema and the first derivative	7, 8, 9, 10, 11, 12, 13, 14, 15	
Concavity and the second derivative	16, 17, 18, 19, 20	
Extrema and the second derivative	21	
Concavity and the second derivative	22, 23, 24	
Optimization	25, 26, 27, 28, 29	
Differentials	30, 31, 32, 33, 34, 35	

Integration

Test Yourself

6.1 Basic Integration Rules

Up to this point, you have been given a differentiable function on some interval and you were asked to determine its derivative. In this chapter, you will be given a derivative and you will be asked to determine the original function.

Definition: A function F is called an *antiderivative* of the function f, in the variable x, if and only if $F'(x) = f(x)$ for every x in the domain of f.

Integration is the process of finding antiderivatives. In the equation $\int f(x)\, dx = F(x) + C$, the symbol "$\int$" represents integration, f is the function being integrated, F is an antiderivative of f, and C represents an arbitrary constant.

The following are some of the most basic integration rules:

1. $\int k\, dx = kx + C$ (Constant Rule)

2. $\int k \cdot f(x)\, dx = k \int f(x)\, dx$ (Constant Times a Function Rule)

3. $\int [f(x) \pm g(x)]\, dx = \int f(x)\, dx \pm \int g(x)\, dx$ (Sum Rule)

4. $\int x^n dx = \dfrac{x^{n+1}}{n+1} + C$, $n \neq -1$ (Power Rule)

5. $\int kx^n\, dx = \dfrac{kx^{n+1}}{n+1} + C$, $n \neq -1$ (Constant Times Power Rule)

In Exercises 1-7, integrate and check your results by differentiation.

1. $\int (2x + 3)\, dx$

2. $\int 2\sqrt{z}\, dz$

3. $\int \left(\dfrac{4}{x^3} - \dfrac{1}{x^4} \right) dx$

4. $\int \dfrac{2x - 3x^2 + \sqrt{x}}{x^{1/3}}\, dx$

5. $\int (3x + 4)(x - 3)\, dx$

6. $\int \dfrac{t^3 + 2t^4}{2t^2}\, dt$

7. $\int y^{2/3}(2y + 1)^2\, dy$

6.2 The General Power Rule

The expressions x^3, $x^{3/2}$, and x^{-4} are powers of the *variable* x. However, the expressions $(2x - 1)^3$, $(1 - 5x)^{1/2}$, and $(2x^3 - 1)^5$ are powers of *functions* of x.

6. The General Power Rule: Let u(x) be a differentiable function of x. Then, $\int u^n\, du = \dfrac{u^{n+1}}{n+1} + C$, $n \neq -1$.

If u is a differentiable function of x, then $du = u'(x)\, dx$. The General Power Rule can now be symbolized as

$$\int u^n du = \int u^n\, u'(x)\, dx = \frac{u^{n+1}}{n+1} + C, n \neq -1$$

Note: The expression $u'(x)$ must be a *factor* of the integrand in order to use the General Power Rule.

In Exercises 8-14, integrate and check your results by differentiation.

8. $\int 2(5 + 2x)^3 \, dx$

9. $\int x(x^2 - 1)^5 \, dx$

10. $\int \sqrt{3y - 1} \, dy$

11. $\int 2t^2 \sqrt{4 - t^3} \, dt$

12. $\int \dfrac{-6x}{(1 - 3x^2)^2} \, dx$

13. $\int (3y^2 + 1)(y^3 + y)^5 \, dy$

14. $\int \left(1 + \dfrac{1}{s}\right)^4 \left(\dfrac{1}{s^2}\right) ds$

6.3 Marginal Analysis Revisited

If f is a differentiable function, then f′ is a unique function. However, if we are given g′, then g will be one of infinitely many antiderivatives, depending upon the value of the constant of integration.

15. Given $f'(x) = 2x - x^2$, determine a simplified expression for $f(x)$ satisfying the condition that $f(1) = -2$.

16. Suppose that the marginal cost of producing x units of a product is given by $C'(x) = 25 - 0.05x$. If it costs \$40 to produce 1 unit, what is the total cost of producing 100 units?

17. Suppose that the marginal revenue from the production and sale of x units of a product is given by $R'(x) = 1200x - 0.06x^2$. If the revenue from the production and sale of 5 units is \$14,997.50, what is the total revenue from the production and sale of 300 units?

18. Suppose that the marginal profit from the production and sale of x units of a product is given by $P'(x) = 0.04x - 40$. If the profit from the production and sale of 10 units is \$3602, what is the total profit from the production and sale of 500 units?

19. Suppose that the marginal average cost per unit of production is given by $\overline{C}'(x) = \dfrac{-48}{x^2}$. If the average cost per unit at a production level of 2 units is \$45, what is the average cost per unit at a production level of 200 units?

20. Suppose that the marginal average revenue per unit of production and sale is given by $\overline{R}'(x) = \dfrac{-1}{15}$. If the average revenue per unit at a production and sale level of 10 units is \$20, what is the average revenue per unit at a production and sale level of 100 units?

6.4 The Definite Integral

Definition: The symbol $\displaystyle\int_a^b f(x) \, dx$ is called the **definite integral from a to b.** The expression f(x) is called the **integrand,** a is the **lower limit of integration,** and b is the **upper limit of integration.**

In Section 6.1, we noted that the indefinite integral $\int f(x) \, dx$ is a family of functions. A definite integral, however, is a number. To evaluate a definite integral, we use the Fundamental Theorem of Calculus.

The Fundamental Theorem of Calculus: If f is a continuous function on the interval [a, b], then

$$\int_a^b f(x) \, dx = F(b) - F(a)$$

where F is any function such that $F'(x) = f(x)$ for all x in [a, b].

In Exercises 21-31, evaluate the definite integrals.

21. $\displaystyle\int_0^1 3x \, dx$

22. $\displaystyle\int_{-2}^2 (y^2 - 1) \, dy$

23. $\displaystyle\int_0^4 (\sqrt{t} - 4) \, dt$

24. $\displaystyle\int_{-3}^{0} p(p^2 + 1)^4 \, dp$

25. $\displaystyle\int_{-1}^{3} (2x^3 - 3x^2 + 7) \, dx$

26. $\displaystyle\int_{2}^{5} \frac{s^3 - 1}{s^2} \, ds$

27. $\displaystyle\int_{-5}^{5} 5 \, dq$

28. $\displaystyle\int_{-1}^{1} x^2 \sqrt{1 - x^3} \, dx$

29. $\displaystyle\int_{0}^{4} \frac{1}{\sqrt{3x + 2}} \, dx$

30. $\displaystyle\int_{1}^{9} \sqrt{t}(2 - t) \, dt$

31. $\displaystyle\int_{0}^{2} y\sqrt{4 - y^2} \, dy$

6.5 Area of a Region Between Two Curves

Let f and g be two continuous functions on the interval [a, b] such that $g(x) \le f(x)$ for all x in [a, b]. Then, the area of the region in the plane bounded by the graphs of $y = f(x)$, $y = g(x)$, $x = a$, and $x = b$ is given by

$$A = \int_{a}^{b} [f(x) - g(x)] \, dx \, .$$

Area is measured in square units.

In Exercises 32-38, determine the area of the region of the plane bounded by the graphs of the given equations.

32. $y = x^2 + 1$, $y = -2$, $x = 0$, and $x = 3$

33. $y = 3 - x^2$, and $y = x + 1$

34. $y = x^3$, and $y = 4x$

35. $y = x^2 - 4$, and $y = -x^2 + 4$

36. $y = x^2 + 4x + 1$, and $y = 4x + 5$

37. $y = 3 + 2x - x^2$, and $y = x^2 - 4x - 5$

38. $y = x^3 + 2$, $y = -2$, $x = -1$, and $x = 2$

✓ Check Yourself

1. $\displaystyle\int (2x + 3) \, dx = \int 2x \, dx + \int 3 \, dx$ (Rule 3)

$\displaystyle = \frac{2x^2}{2} + \int 3 \, dx$ (Rule 5)

$= x^2 + 3x + C$ (Rule 1)

Therefore, $\int (2x + 3) \, dx = x^2 + 3x + C$ for all real values of x.

Check: $\dfrac{d}{dx} (x^2 + 3x + C) = 2x + 3 + 0$

$= 2x + 3$ for all real values of x. √ **(Basic integration rules)**

2. $\int 2\sqrt{z}\ dz = \int 2z^{1/2}\ dz$ (Rewrite)

$$= 2\left(\frac{z^{3/2}}{3/2}\right) + C \qquad\qquad\text{(Rule 5)}$$

$$= \frac{4}{3}z^{3/2} + C$$

Therefore, $\int 2\sqrt{z}\ dz = \frac{4}{3}z^{3/2} + C$ for all real values of $z \geq 0$.

Check: $\dfrac{d}{dz}\left(\dfrac{4}{3}z^{3/2} + C\right) = \left(\dfrac{4}{3}\right)\left(\dfrac{3}{2}\right)z^{1/2} + 0 = 2z^{1/2} = 2\sqrt{z}$ for all real values of $z \geq 0$. $\sqrt{}$ **(Basic integration rules)**

3. $\int\left(\dfrac{4}{x^3} - \dfrac{1}{x^4}\right)dx = \int 4x^{-3}\ dx - \int x^{-4}\ dx$ (Rule 3)

$$= 4\left(\frac{x^{-2}}{-2}\right) - \int x^{-4}\ dx \qquad\text{(Rule 5)}$$

$$= -2x^{-2} - \frac{x^{-3}}{-3} + C \qquad\text{(Rule 4)}$$

$$= \frac{-2}{x^2} + \frac{1}{3x^3} + C$$

Therefore, $\int\left(\dfrac{4}{x^3} - \dfrac{1}{x^4}\right)dx = \dfrac{-2}{x^2} + \dfrac{1}{3x^3} + C$ for all real values of $x \neq 0$.

Check: $\dfrac{d}{dx}\left(\dfrac{-2}{x^2} + \dfrac{1}{3x^3} + C\right) = \dfrac{d}{dx}(-2x^{-2} + \dfrac{1}{3}x^{-3} + C)$

$$= (-2)(-2)x^{-3} + \frac{1}{3}(-3)x^{-4} + 0$$

$$= 4x^{-3} - x^{-4}$$

$$= \frac{4}{x^3} - \frac{1}{x^4}\text{ for all real values of }x \neq 0.\ \sqrt{}\ \textbf{(Basic integration rules)}$$

4. $\int\dfrac{2x - 3x^2 + \sqrt{x}}{x^{1/3}}\ dx = \int(2x^{2/3} - 3x^{5/3} + x^{1/6})\ dx$ (Rewrite)

$$= \int 2x^{2/3}\ dx - \int 3x^{5/3}\ dx + \int x^{1/6}\ dx \qquad\text{(Rule 3)}$$

$$= 2\left(\frac{x^{5/3}}{5/3}\right) - 3\left(\frac{x^{8/3}}{8/3}\right) + \frac{x^{7/6}}{7/6} + C \qquad\text{(Rule 5)}$$

$$= \frac{6}{5}x^{5/3} - \frac{9}{8}x^{8/3} + \frac{6}{7}x^{7/6} + C\text{ for all real values of }x > 0.$$

Check: $\dfrac{d}{dx}\left(\dfrac{6}{5}x^{5/3} - \dfrac{9}{8}x^{8/3} + \dfrac{6}{7}x^{7/6} + C\right) = \left(\dfrac{6}{5}\right)\left(\dfrac{5}{3}\right)x^{2/3} - \left(\dfrac{9}{8}\right)\left(\dfrac{8}{3}\right)x^{5/3} + \left(\dfrac{6}{7}\right)\left(\dfrac{7}{6}\right)x^{1/6} + 0$

$$= 2x^{2/3} - 3x^{5/3} + x^{1/6}\text{ for all real values of }x > 0.\ \sqrt{}\ \textbf{(Basic}$$

integration rules)

5. $\int (3x + 4)(x - 3)\, dx = \int (3x^2 - 5x - 12)\, dx$ (Rewrite)

$$= \int 3x^2\, dx - \int 5x\, dx - \int 12\, dx \quad\text{(Rule 3)}$$

$$= x^3 - \frac{5}{2}x^2 - 12x + C \text{ for all real values of x.} \quad\text{(Rules 1 \& 5)}$$

Check: $\dfrac{d}{dx}(x^3 - \dfrac{5}{2}x^2 - 12x + C) = 3x^2 - \dfrac{5}{2}(2x) - 12 + 0$

$$= 3x^2 - 5x - 12$$

$$= (3x + 4)(x - 3) \text{ for all real values of x.} \;\checkmark \textbf{ (Basic integration rules)}$$

6. $\int \dfrac{t^3 + 2t^4}{2t^2}\, dt = \dfrac{1}{2}\int \dfrac{t^3 + 2t^4}{t^2}\, dt$ (Rule 2)

$$= \frac{1}{2}\int (t + 2t^2)\, dt \qquad\text{(Rewrite)}$$

$$= \frac{1}{2}\int t\, dt + \frac{1}{2}\int 2t^2\, dt \qquad\text{(Rule 3)}$$

$$= \frac{1}{4}t^2 + \left(\frac{1}{2}\right)\left(\frac{2t^3}{3}\right) + C \qquad\text{(Rules 4 \& 5)}$$

$$= \frac{1}{4}t^2 + \frac{1}{3}t^3 + C \text{ for all real values of } t \neq 0.$$

Check: $\dfrac{d}{dt}\left(\dfrac{1}{4}t^2 + \dfrac{1}{3}t^3 + C\right) = \dfrac{1}{4}(2t) + \dfrac{1}{3}(3t^2) + 0\}$

$$= \frac{1}{2}t + t^2 \text{ for all real values of } t \neq 0. \;\;\checkmark \textbf{ (Basic integration rules)}$$

7. $\int y^{2/3}(2y + 1)^2\, dy = \int y^{2/3}(4y^2 + 4y + 1)\, dy$ (Rewrite)

$$= \int (4y^{8/3} + 4y^{5/3} + y^{2/3})\, dy \qquad\text{(Rewrite)}$$

$$= \int 4y^{8/3}\, dy + \int 4y^{5/3}\, dy + \int y^{2/3}\, dy \qquad\text{(Rule 3)}$$

$$= \frac{12}{11}y^{11/3} + \frac{3}{2}y^{8/3} + \frac{y^{5/3}}{5/3} + C \text{ (Rules 4 \& 5)}$$

$$= \frac{12}{11}y^{11/3} + \frac{3}{2}y^{8/3} + \frac{3}{5}y^{5/3} + C \text{ for all real values of y.}$$

Check: $\dfrac{d}{dy}\left(\dfrac{12}{11}y^{11/3} + \dfrac{3}{2}y^{8/3} + \dfrac{3}{5}y^{5/3} + C\right)$

$$= \left(\frac{12}{11}\right)\left(\frac{11}{3}\right)y^{8/3} + \left(\frac{3}{2}\right)\left(\frac{8}{3}\right)y^{5/3} + \left(\frac{3}{5}\right)\left(\frac{5}{3}\right)y^{2/3} + 0$$

$$= 4y^{8/3} + 4y^{5/3} + y^{2/3}$$

$$= y^{2/3}(4y^2 + 4y + 1)$$

$$= y^{2/3}(2y + 1)^2 \text{ for all real values of y.} \;\;\checkmark \textbf{ (Basic integration rules)}$$

8. $\int 2(5 + 2x)^3 \, dx$ for all real values of x.

   Using The General Power Rule, we let $u(x) = 5 + 2x$ and $n = 3$.

   Hence, $u'(x) = \dfrac{d}{dx}(5 + 2x) = 2$. We have

   $$\int (5 + 2x)^3 \, (2) \, dx = \int u^n \, u'(x) \, dx = \frac{u^{n+1}}{n+1} + C$$

   $$= \frac{(5 + 2x)^4}{4} + C \text{ for all real values of x.}$$

   Check: $\dfrac{d}{dx}\left(\dfrac{(5 + 2x)^4}{4} + C\right) = \dfrac{1}{4}(4)(5 + 2x)^3(2) + 0$

   $$= 2(5 + 2x)^3 \text{ for all real values of x.} \quad \sqrt{} \text{ (The General Power Rule)}$$

9. $\int x(x^2 - 1)^5 \, dx$ for all real values of x.

   Using The General Power Rule, we let $u(x) = x^2 - 1$ and $n = 5$.

   Hence, $u'(x) = \dfrac{d}{dx}(x^2 - 1) = 2x$. We have

   $$\int x(x^2 - 1)^5 \, dx = \int (x^2 - 1)^5 \, x \, dx = \int u^n \, \frac{1}{2} u'(x) \, dx$$

   $$= \frac{1}{2}\int u^n \, u'(x) \, dx = \frac{1}{2}\left(\frac{u^{n+1}}{n+1}\right) + C$$

   $$= \frac{1}{2}\frac{(x^2 - 1)^6}{6} + C = \frac{1}{12}(x^2 - 1)^6 + C \text{ for all real values of x.}$$

   Check: $\dfrac{d}{dx}\left[\dfrac{1}{12}(x^2 - 1)^6 + C\right] = \dfrac{1}{12}(6)(x^2 - 1)^5(2x) + 0$

   $$= x(x^2 - 1)^5 \text{ for all real values of x.} \quad \sqrt{} \text{ (The General Power Rule)}$$

10. $\int \sqrt{3y - 1} \, dy$ for all $y \geq 1/3$.

    Using The General Power Rule, we let $u(y) = 3y - 1$ and $n = 1/2$.

    Hence, $u'(y) = \dfrac{d}{dy}(3y - 1) = 3$. We have

    $$\int \sqrt{3y - 1} \, dy = \int u^n\left(\frac{1}{3}u'(y)\right) dy = \frac{1}{3}\int u^n \, u'(y) \, dy$$

    $$= \frac{1}{3}\left(\frac{u^{n+1}}{n+1}\right) + C = \frac{1}{3}\frac{(3y - 1)^{3/2}}{3/2} + C$$

    $$= \frac{2}{9}(3y - 1)^{3/2} + C \text{ for all real values of } y \geq 1/3.$$

    Check: $\dfrac{d}{dy}\left[\dfrac{2}{9}(3y - 1)^{3/2} + C\right] = \left(\dfrac{2}{9}\right)\left(\dfrac{3}{2}\right)(3y - 1)^{1/2}(3) + 0$

    $$= \sqrt{3y - 1} \text{ for all real values of } y \geq 1/3. \quad \sqrt{} \text{ (The General Power Rule)}$$

11. $\int 2t^2 \sqrt{4 - t^3}$ dt for all real values of t $\le \sqrt[3]{4}$

Using The General Power Rule, we let $u(t) = 4 - t^3$ and n = 1/2.

Hence, $u'(t) = \dfrac{d}{dt}(4 - t^3) = -3t^2$. We have

$\int 2t^2\sqrt{4 - t^3}\ dt = 2\int t^2\sqrt{4 - t^3}\ dt = 2\int \sqrt{4 - t^3}\ (t^2)dt$

$\qquad = 2\int u^n \left[\dfrac{-1}{3}u\,'(t)\right]dt = \dfrac{-2}{3}\int u^n\, u\,'(t)\ dt =$

$\qquad = \dfrac{-2}{3}\,\dfrac{u^{n+1}}{n+1} + C = \dfrac{-2}{3}\,\dfrac{(4 - t^3)^{3/2}}{3/2} + C$

$\qquad = \dfrac{-4}{9}(4 - t^3)^{3/2} + C$ for all real values values of t $\le \sqrt[3]{4}$.

Check: $\dfrac{d}{dt}\left[\dfrac{-4}{9}(4 - t^3)^{3/2} + C\right] = \left(\dfrac{-4}{9}\right)\left(\dfrac{3}{2}\right)(4 - t^3)^{1/2}(-3t^2) + 0$

$\qquad\qquad = 2t^2(4 - t^3)^{1/2}$

$\qquad\qquad = 2t^2\sqrt{4 - t^3}$ for all real values of t $\le \sqrt[3]{4}$. $\sqrt{}$ **(The General Power Rule)**

12. $\int \dfrac{-6x}{(1 - 3x^2)^2}$ dx for all real values of $x \ne \pm 1/\sqrt{3}$.

Using The General Power Rule, we let $u(x) = 1 - 3x^2$ and n = -2. Hence, $u'(x) = \dfrac{d}{dx}(1 - 3x^2) = -6x$. We have

$\int \dfrac{-6x}{(1 - 3x^2)^2}\ dx = \int (1 - 3x^2)^{-2}(-6x)\ dx = \int u^n\, u'(x)\ dx$

$\qquad = \dfrac{u^{n+1}}{n+1} + C = \dfrac{(1 - 3x^2)^{-1}}{-1} + C$

$\qquad = \dfrac{-1}{1 - 3x^2} + C$ for all real values of $x \ne \pm 1/\sqrt{3}$.

Check: $\dfrac{d}{dx}\left(\dfrac{-1}{1 - 3x^2} + C\right) = \dfrac{d}{dx}[-(1 - 3x^2)^{-1}]$

$\qquad\qquad = (-1)(-1)(1 - 3x^2)^{-2}(-6x) + 0 = \dfrac{-6x}{(1 - 3x^2)^2}$ for all real values of $x \ne \pm 1/\sqrt{3}$. $\sqrt{}$

(The General Power Rule)

13. $\int (3y^2 + 1)(y^3 + y)^5$ dy for all real values of y.

Using The General Power Rule, we let $u(y) = y^3 + y$ and n = 5. Hence, $u'(y) = \dfrac{d}{dy}(y^3 + y) = 3y^2 + 1$. We have

$\int (3y^2 + 1)(y^3 + y)^5\ dy = \int (y^3 + y)^5(3y^2 + 1)\ dy$

$$= \int u^n \, u'(y) \, dy = \frac{u^{n+1}}{n+1} + C = \frac{(y^3 + y)^6}{6} + C \text{ for all real values of y.}$$

Check: $\dfrac{d}{dy}\left[\dfrac{(y^3 + y)^6}{6} + C\right] = \dfrac{1}{6}(6)(y^3 + y)^5(3y^2 + 1) + 0$

$$= (y^3 + y)^5(3y^2 + 1) \text{ for all real values of y. } \sqrt{} \text{ (\textbf{The General Power Rule})}$$

14. $\displaystyle\int\left(1 + \frac{1}{s}\right)^4\left(\frac{1}{s^2}\right) ds$ for all real values of $s \neq 0$.

Using The general power rule, we let $u(s) = 1 + \dfrac{1}{s}$ and $n = 4$.

Hence, $u'(s) = \dfrac{d}{ds}\left(1 + \dfrac{1}{s}\right) = 0 - \dfrac{1}{s^2} = \dfrac{-1}{s^2}$. We have

$$\int\left(1 + \frac{1}{s}\right)^4\left(\frac{1}{s^2}\right) ds = \int u^n \, (-1) \, u'(s) \, ds = -\int u^n \, u'(s) \, ds$$

$$= (-1)\frac{u^{n+1}}{n+1} + C = (-1)\frac{[1 + (1/s)]^5}{5} + C$$

$$= \frac{-1}{5}\left(1 + \frac{1}{s}\right)^5 + C \text{ for all real values of } s \neq 0.$$

Check: $\dfrac{d}{ds}\left[\dfrac{-1}{5}\left(1 + \dfrac{1}{s}\right)^5\right] = \dfrac{-1}{5}(5)\left(1 + \dfrac{1}{s}\right)^4\left(\dfrac{-1}{s^2}\right) + 0$

$$= \left(1 + \frac{1}{s}\right)^4\left(\frac{1}{s^2}\right) \text{ for all real values of } s \neq 0. \sqrt{} \text{ (\textbf{The General Power Rule})}$$

15. Since $f'(x) = 2x - x^2$, then

$$f(x) = \int (2x - x^2) \, dx = \frac{2x^2}{2} - \frac{x^3}{3} + C = x^2 - \frac{x^3}{3} + C$$

To determine the particular antiderivative satisfying the condition that $f(1) = -2$, we must solve for C as follows:

$$f(x) = x^2 - \frac{x^3}{3} + C$$

$$f(1) = (1) - \frac{(1)^3}{3} + C = -2$$

$$1 - \frac{1}{3} + C = -2$$

$$C = \frac{-8}{3}$$

Therefore, the required antiderivative is $f(x) = x^2 - \dfrac{x^3}{3} - \dfrac{8}{3}$. **(Marginal analysis revisited)**

16. Since $C'(x) = 25 - 0.05x$, then

$$C(x) = \int (25 - 0.05x)\, dx = 25x - \frac{0.05x^2}{2} + C.$$

But, when $x = 1$, $C(x) = 40$. Hence,

$$C(1) = 25(1) - \frac{0.05(1)^2}{2} + C = 40$$

$$25 - 0.025 + C = 40$$

$$C = 15.025$$

Therefore, $C(x) = 25x - \frac{0.05x^2}{2} + 15.025$. Hence,

$$C(100) = 25(100) - \frac{0.05(100)^2}{2} + 15.025 \text{ (in dollars)}$$

$$= 2500 - 250 + 15.025$$

$$= 2265.025$$

$$= \$2265.03 \text{ (\textbf{Marginal analysis revisited})}$$

17. Since $R'(x) = 1200x - 0.06x^2$, then

$$R(x) = \int (1200x - 0.06x^2)\, dx = \frac{1200x^2}{2} - \frac{0.06x^3}{3} + C$$

$$= 600x^2 - 0.02x^3 + C$$

But, when $x = 5$, $R(x) = 14{,}997.50$. Hence,

$$R(5) = 600(5)^2 - 0.02(5)^3 + C = 14{,}997.50$$

$$600(25) - 0.02(125) + C = 14{,}997.50$$

$$15{,}000 - 2.5 + C = 14{,}997.50$$

$$C = 0$$

Therefore, $R(x) = 600x^2 - 0.02x^3$. Hence,

$$R(300) = 600(300)^2 - 0.02(300)^3 \text{ (in dollars)}$$

$$= 600(90{,}000) - 0.02(27{,}000{,}000)$$

$$= 54{,}000{,}000 - 540{,}000$$

$$= \$53{,}460{,}000 \text{ (\textbf{Marginal analysis revisited})}$$

18. Since $P'(x) = 0.04x - 40$, then

$$P(x) = \int (0.04x - 40)\, dx = \frac{0.04x^2}{2} - 40x + C$$

$$= 0.02x^2 - 40x + C$$

But, when x = 10, P(x) = 3602. Hence,

$$P(10) = 0.02(10)^2 - 40(10) + C = 3602 \text{ (in dollars)}$$

$$0.02(100) - 400 + C = 3602$$

$$2 - 400 + C = 3602$$

$$C = 4000$$

Therefore, $P(x) = 0.02x^2 - 40x + 4000$. Hence,

$$P(500) = 0.02(500)^2 - 40(500) + 4000 \text{ (in dollars)}$$

$$= 0.02(250,000) - 20,000 + 4000$$

$$= 5000 - 20,000 + 4000$$

$$= -\$11,000 \text{ (a loss)} \textbf{ (Marginal analysis revisited)}$$

19. Since $\bar{C}'(x) = \dfrac{-48}{x^2}$, then

$$\bar{C}(x) = \int \left(\frac{-48}{x^2}\right) dx = \int (-48x^{-2}) \, dx = \frac{-48x^{-1}}{-1} + C$$

$$= \frac{48}{x} + C$$

But, when x = 2, $\bar{C}(x) = 45$. Hence,

$$\bar{C}(2) = \frac{48}{2} + C = 45$$

$$24 + C = 45$$

$$C = 21$$

Therefore, $\bar{C}(200) = \dfrac{48}{200} + 21 \text{ (in dollars)}$

$$= 0.24 + 21$$

$$= \$21.24 \textbf{ (Marginal analysis revisited)}$$

20. Since $\bar{R}'(x) = \dfrac{-1}{15}$, then

$$\bar{R}(x) = \int \left(\frac{-1}{15}\right) dx = \frac{-1}{15}x + C$$

But, when x = 10, $\bar{R}(x) = 20$. Hence,

$$\bar{R}(10) = \frac{-1}{15}(10) + C = 20$$

$$\frac{-2}{3} + C = 20$$

$$C = \frac{62}{3}$$

Therefore, $\bar{R}(x) = \dfrac{-x}{15} + \dfrac{62}{3}$. Hence,

$$\bar{R}(100) = \dfrac{-100}{15} + \dfrac{62}{3} \text{ (in dollars)}$$

$$= \dfrac{-20}{3} + \dfrac{62}{3}$$

$$= \$14 \textbf{ (Marginal analysis revisited)}$$

21. $\displaystyle\int_0^1 3x \, dx = \left[\dfrac{3x^2}{2}\right]_0^1 = \dfrac{3}{2}[(1)^2 - (0)^2] = \dfrac{3}{2}(1) = \dfrac{3}{2}$ **(The definite integral)**

22. $\displaystyle\int_{-2}^2 (y^2 - 1) \, dy = \left[\dfrac{y^3}{3} - y\right]_{-2}^2 = \left[\dfrac{8}{3} - 2\right] - \left[\dfrac{-8}{3} + 2\right]$

$$= \dfrac{16}{3} - 4 = \dfrac{4}{3} \textbf{ (The definite integral)}$$

23. $\displaystyle\int_0^4 (\sqrt{t} - 4) \, dt = \left[\dfrac{t^{3/2}}{3/2} - 4t\right]_0^4$

$$= \left[\dfrac{4^{3/2}}{3/2} - 4(4)\right] - \left[\dfrac{0^{3/2}}{3/2} - 4(0)\right]$$

$$= \left[\dfrac{8}{3/2} - 16\right] - (0 - 0) = \dfrac{16}{3} - 16 = \dfrac{-32}{3} \textbf{ (The definite integral)}$$

24. $\displaystyle\int_{-3}^0 p(p^2 + 1)^4 \, dp = \left(\dfrac{1}{2}\right) \int_{-3}^0 (p^2 + 1)^4 (2p) \, dp$

$$= \dfrac{1}{2}\left[\dfrac{(p^2 + 1)^5}{5}\right]_{-3}^0 = \dfrac{1}{10}\left[(p^2 + 1)^5\right]_{-3}^0$$

$$= \dfrac{(0 + 1)^5}{10} - \dfrac{(9 + 1)^5}{10} = \dfrac{1}{10} - \dfrac{10^5}{10}$$

$$= 0.1 - 10{,}000 = -9999.9 \textbf{ (The definite integral)}$$

25. $\displaystyle\int_{-1}^3 (2x^3 - 3x^2 + 7) \, dx = \left[\dfrac{2x^4}{4} - \dfrac{3x^3}{3} + 7x\right]_{-1}^3$

$$= \left[\dfrac{x^4}{2} - x^3 + 7x\right]_{-1}^3 = \left[\dfrac{81}{2} - 27 + 21\right] - \left[\dfrac{1}{2} + 1 - 7\right]$$

$$= \dfrac{81}{2} - 6 + \dfrac{11}{2} = 40 \textbf{ (The definite integral)}$$

26. $\displaystyle\int_{2}^{5} \frac{s^3-1}{s^2}\, ds = \int_{2}^{5}\left(s - \frac{1}{s^2}\right)ds = \left[\frac{s^2}{2} - \frac{s^{-1}}{-1}\right]_{2}^{5}$

$\qquad = \left[\frac{s^2}{2} + \frac{1}{s}\right]_{2}^{5} = \left[\frac{25}{2} + \frac{1}{5}\right] - \left[\frac{4}{2} + \frac{1}{2}\right] = \left[\frac{127}{10}\right] - \left[2 + \frac{1}{2}\right]$

$\qquad = \dfrac{127}{10} - \dfrac{5}{2} = 10.2$ **(The definite integral)**

27. $\displaystyle\int_{-5}^{5} 5\, dq = [5q]_{-5}^{5} = 5[5 - (-5)] = 5(10) = 50$ **(The definite integral)**

28. $\displaystyle\int_{-1}^{1} x^2\sqrt{1-x^3}\, dx = \frac{-1}{3}\int_{-1}^{1}\sqrt{1-x^3}(-3x^2)dx$

$\qquad = \dfrac{-1}{3}\left[\dfrac{(1-x^3)^{3/2}}{3/2}\right]_{-1}^{1} = \dfrac{-2}{9}\left[(1-x^3)^{3/2}\right]_{-1}^{1}$

$\qquad = \dfrac{-2}{9}\left[(1-1)^{3/2} - (1+1)^{3/2}\right] = \dfrac{-2}{9}(0 - 2^{3/2}) = \left(\dfrac{-2}{9}\right)(-2\sqrt{2})$

$\qquad = \dfrac{4\sqrt{2}}{9}$ **(The definite integral)**

29. $\displaystyle\int_{0}^{4} \frac{1}{\sqrt{3x+2}}\, dx = \int_{0}^{4}(3x+2)^{-1/2}\, dx = \frac{1}{3}\int_{0}^{4}(3x+2)^{-1/2}(3)\, dx$

$\qquad = \dfrac{1}{3}\left[\dfrac{(3x+2)^{1/2}}{1/2}\right]_{0}^{4} = \dfrac{2}{3}\left[\sqrt{3x+2}\ \right]_{0}^{4}$

$\qquad = \dfrac{2}{3}[\sqrt{12+2} - \sqrt{0+2}] = \dfrac{2}{3}[\sqrt{14} - \sqrt{2}]$ **(The definite integral)**

30. $\displaystyle\int_{1}^{9} \sqrt{t}\,(2-t)\, dt = \int_{1}^{9}(2\sqrt{t} - t^{3/2})\, dt = \left[\frac{2t^{3/2}}{3/2} - \frac{t^{5/2}}{5/2}\right]_{1}^{9}$

$\qquad = \left[\dfrac{4t^{3/2}}{3} - \dfrac{2t^{5/2}}{5}\right]_{1}^{9} = \left[\dfrac{4}{3}(9)^{3/2} - \dfrac{2}{5}(9)^{5/2}\right] - \left[\dfrac{4}{3}(1)^{3/2} - \dfrac{2}{5}(1)^{5/2}\right]$

$\qquad = \left[\dfrac{4}{3}(27) - \dfrac{2}{5}(243)\right] - \left[\dfrac{4}{3} - \dfrac{2}{5}\right] = \dfrac{108}{3} - \dfrac{486}{5} - \dfrac{4}{3} + \dfrac{2}{5} = \dfrac{104}{3} - \dfrac{484}{5}$

$\qquad = \dfrac{-932}{15}$ **(The definite integral)**

31. $\int_0^2 y\sqrt{4-y^2}\,dy = \frac{-1}{2}\int_0^2 y\sqrt{4-y^2}(-2y)\,dy = \frac{-1}{2}\left[\frac{(4-y^2)^{3/2}}{3/2}\right]_0^2$

$\qquad = \frac{-1}{3}\left[(4-y^2)^{3/2}\right]_0^2 = \frac{-1}{3}\left[(4-4)^{3/2} - (4-0)^{3/2}\right]$

$\qquad = \frac{-1}{3}(0-8) = \frac{8}{3}$ **(The definite integral)**

32. $y = x^2 + 1$, $y = -2$, $x = 0$, and $x = 3$

A sketch of the region is given below. Observe that, on the interval [0, 3], $-2 \le x^2 + 1$. Hence,

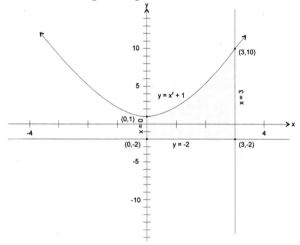

$A = \int_0^3 [(x^2 + 1) - (-2)]\,dx =$

$\qquad = \int_0^3 (x^2 + 3)\,dx = \left[\frac{x^3}{3} + 3x\right]_0^3 =$

$\qquad = (9 + 9) - (0 + 0) = 9 + 9$

$\qquad = 18$ sq units **(Area between two curves)**

33. $y = 3 - x^2$, and $y = x + 1$

A sketch of the region is given on the following page. The points of intersection of the two curves are $(-2, -1)$ and $(1, 2)$. Hence, the region is bounded between $x = -2$ and $x = 1$. We have

$A = \int_{-2}^1 [(3 - x^2) - (x + 1)]\,dx =$

$\qquad = \int_{-2}^1 (-x^2 - x + 2)\,dx = \left[\frac{-x^3}{3} - \frac{x^2}{2} + 2x\right]_{-2}^1$

$\qquad = \left[\frac{-1}{3} - \frac{1}{2} + 2\right] - \left[\frac{8}{3} - 2 - 4\right]$

$\qquad = \frac{7}{6} + \frac{10}{3} = \frac{27}{6} = 4.5$ sq units **(Area between two curves)**

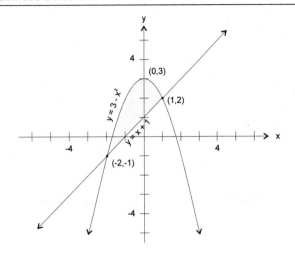

34. $y = x^3$, and $y = 4x$

A sketch of the region is given below. The points of intersection of the two curves are $(-2, -8)$ and $(2, 8)$. Observe that, on the interval $[-2, 0]$, $4x \le x^3$, but that on the interval $[0, 2]$, $x^3 \le 4x$. Hence, the area of the region will have to be found in two parts. We have

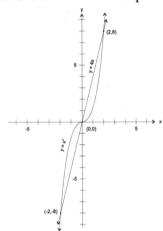

$$A = \int_{-2}^{0} (x^3 - 4x)\, dx + \int_{-2}^{0} (4x - x^3)\, dx$$

$$= \left[\frac{x^4}{4} - \frac{4x^2}{2} \right]_{-2}^{0} + \left[\frac{4x^2}{2} - \frac{x^4}{4} \right]_{0}^{2}$$

$$= \left[\frac{x^4}{4} - 2x^2 \right]_{-2}^{0} + \left[2x^2 - \frac{x^4}{4} \right]_{0}^{2} = [(0 - 0) - (4 - 8)] + [(8 - 4) - 0 - 0)]$$

$$= [0 - (-4)] + (4 - 0) = 4 + 4 = 8 \text{ sq units} \quad \textbf{(Area between two curves)}$$

35. $y = x^2 - 4$, and $y = -x^2 + 4$

A sketch of the region is given on the following page. The points of intersection of the two curves are $(-2, 0)$ and $(2, 0)$. Everywhere on $[-2, 2]$, $x^2 - 4 \le -x^2 + 4$. Hence,

$$A = \int_{-2}^{2} [(-x^2 + 4) - (x^2 - 4)]\, dx$$

$$= \int_{-2}^{2} (-2x^2 + 8)\, dx = \left[\frac{-2x^3}{3} + 8x \right]_{-2}^{2}$$

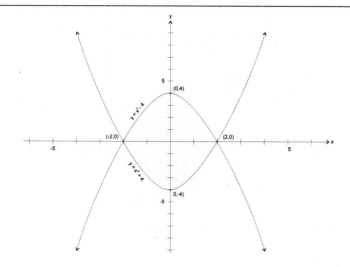

$$= \left[\frac{-16}{3} + 16\right] - \left[\frac{16}{3} - 16\right]$$

$$= \frac{32}{3} + \frac{32}{3} = \frac{64}{3} \text{ sq units} \quad \textbf{(Area between two curves)}$$

36. $y = x^2 + 4x + 1$, and $y = 4x + 5$

A sketch of the region is given below. The points of intersection of the two curves are $(-2, -3)$ and $(2, 13)$. Everywhere on $[-2, 2]$, $x^2 + 4x + 1 \leq 4x + 5$. We have

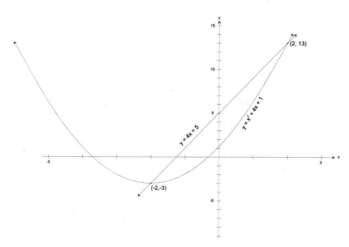

$$A = \int_{-2}^{2} [(4x + 5) - (x^2 + 4x + 1)] \, dx$$

$$= \int_{-2}^{2} (-x^2 + 4) \, dx = \left[\frac{-x^3}{3} + 4x\right]_{-2}^{2}$$

$$= \left[\frac{-8}{3} + 8\right] - \left[\frac{8}{3} - 8\right]$$

$$= \frac{16}{3} + \frac{16}{3} = \frac{32}{3} \text{ sq units} \quad \textbf{(Area between two curves)}$$

37. $y = 3 + 2x - x^2$, and $y = x^2 - 4x - 5$

A sketch of the region is given below. The points of intersection of the two curves are $(-1, 0)$ and $(4, -5)$. Everywhere on $[-1, 4]$, $x^2 - 4x - 5 \leq 3 + 2x - x^2$. We have

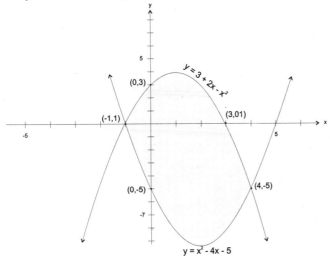

$$A = \int_{-1}^{4} [(3 + 2x - x^2) - (x^2 - 4x - 5)]\, dx$$

$$= \int_{-1}^{4} (-2x^2 + 6x + 8)\, dx = \left[\frac{-2x^3}{3} + 3x^2 + 8x \right]_{-1}^{4}$$

$$= \left[\frac{-128}{3} + 48 + 32 \right] - \left[\frac{2}{3} + 3 - 8 \right] = \left[80 - \frac{128}{3} \right] - \left[\frac{2}{3} - 5 \right]$$

$$= 85 - \frac{130}{3} = \frac{125}{3} \text{ sq units } \textbf{(Area between two curves)}$$

38. $y = x^3 + 2$, $y = -2$, $x = -1$, and $x = 2$

A sketch of the region is given below. Everywhere on $[-1, 2]$, $-2 \le x^3 + 2$. We have

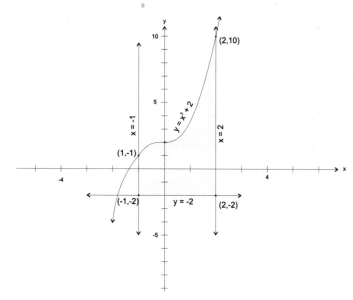

$A = \int_{-1}^{2} [(x^3 + 2) - (-2)] \, dx$

$= \int_{-1}^{2} (x^3 + 4) \, dx = \left[\dfrac{x^4}{4} + 4x \right]_{-1}^{2}$

$= (4 + 8) - (1/4 - 4)$

$= 12 + \dfrac{15}{4} = \dfrac{63}{4}$ sq units **(Area between two curves)**

Grade Yourself

Circle the numbers of the questions you missed, then fill in the total incorrect for each topic. If you answered more than three questions incorrectly, you need to focus on that topic. (If a topic has less than three questions and you had at least one wrong, we suggest you study that topic also. Read your textbook, a review book, or ask your teacher for help.)

Subject: Integration

Topic	Question Numbers	Number Incorrect
Basic integration rules	1, 2, 3, 4, 5, 6, 7	
The General Power Rule	8, 9, 10, 11, 12, 13, 14	
Marginal analysis revisited	15, 16, 17, 18, 19, 20	
The definite integral	21, 22, 23, 24, 25, 26, 27, 28, 29, 30, 31	
Area between two curves	32, 33, 34, 35, 36, 37, 38	

Exponential and Logarithmic Functions

Test Yourself

7.1 Exponential Functions

Definition: If $b > 0$ and $b \neq 1$, then $y = b^x$ defines an **exponential function** with base b.

Observations about exponential functions:

1. The domain of every exponential function is $(-\infty, +\infty)$.

2. The range of every exponential function is $(0, +\infty)$.

3. Every exponential function is one-to-one.

4. Every exponential function has an inverse function.

5. If $b > 1$, the exponential function is increasing (and its graph is rising from left to right).

6. If $0 < b < 1$, the exponential function is decreasing (and its graph is falling from left to right).

7. The graph of every exponential function passes through the point $(0, 1)$ in the plane containing the graph.

8. The horizontal axis of the plane containing the graph of an exponential function is a horizontal asymptote for the graph.

In Exercises 1-7, sketch the graph of the given exponential functions.

1. $y = 3^x$

2. $y = 3^{-x}$

3. $y = -2^x + 1$

4. $y = 4^x + 3$

5. $y = 3^{-x} - 2$

6. $y = \left(\frac{1}{4}\right)^x$

7. $y = \frac{3^x}{5}$

Compound Interest: If P dollars are invested at an annual interest rate r and the interest is compounded n times per year, then the total amount, A, of the investment at the end of t years is given by the formula

$$A = P\left(1 + \frac{r}{n}\right)^{nt}$$

In Exercises 8-10, determine the total amount of investment for the given P, r, n, and t.

8. P = \$1000, r = 8%, n = 2, t = 3 years

9. P = $2000, r = 6%, n = 12, t = 42 months

10. P = $2000, r = 7.5%, n = 12, t = 10 years

Continuous Compound Interest: If P dollars are invested at an annual interest rate r and the interest is compounded continuously, then the total amount, A, of the investment at the end of t years is given by the formula

$$A = Pe^{rt}.$$

In Exercises 11-13, determine the total amount of investment for the given data if interest is compounded continuously.

11. P = $800, r = 7%, t = 1 year

12. P = $900, r = 8%, t = 30 months

13. P = $20,000, r = 10%, t = 6 months

7.2 Differentiation and Integration of Exponential Functions

We have the following rules for differentiation of exponential functions: (b > 0, b ≠ 1, u is a differentiable function of x).

Derivative of e^x: $\dfrac{d}{dx}(e^x) = e^x$ for all real values of x.

Derivative of e^u: $\dfrac{d}{dx}(e^u) = e^u \dfrac{du}{dx}$ for all real values of u.

Derivative of b^x: $\dfrac{d}{dx}(b^x) = b^x \ln(b)$ for all real values of x.

Derivative of b^u: $\dfrac{d}{dx}(b^u) = b^u \ln(b) \dfrac{du}{dx}$ for all real values of u.

In Exercises 14-20, determine a simplified expression for f'(x).

14. $f(x) = 3^x$

15. $f(x) = e^{3x}$

---

16. $f(x) = 50e^{x^2+1}$

17. $f(x) = (2 + e^x)^3$

18. $f(x) = xe^x$

19. $f(x) = \dfrac{2x}{e^x}$

20. $f(x) = (3 - x)5^x$

We have the following rules for integration of exponential functions: (b > 0, b ≠ 1, u is a differentiable function of x).

Integral of e^x: $\int e^x \, dx = e^x + C$ for all real values of x.

Integral of e^u: $\int e^u \, u'(x) \, dx = \int e^u \, du = e^u + c$ for all real values of u.

Integral of b^x: $\int b^x \, dx = \dfrac{b^x}{\ln(b)} + C$ for all real values of x.

Integral of b^u: $\int b^u \, u'(x) \, dx = \int b^u \, du = \dfrac{b^u}{\ln(b)} + C$ for all real values of u.

In Exercises 21-27, integrate:

21. $\int e^{2x} \, dx$

22. $\int xe^{2x^2} \, dx$

23. $\int \dfrac{5^{2/x}}{x^2} \, dx$

24. $\int (e^x + e^{-x})^2 \, dx$

25. $\int e^x(1 - e^x)^4 \, dx$

26. $\int (4^{3x-1} + 2) \, dx$

27. $\int (e^{3x} + e^{5x}) \, dx$

7.3 Logarithmic Functions

Definition: If b > 0 and b ≠ 1, then $y = \log_b(x)$ defines a **logarithmic function** with base b if, and only if, $b^y = x$.

Observations about logarithmic functions:

1. The domain of every logarithmic function is $(0, +\infty)$.

2. The range of every logarithmic function is $(-\infty, +\infty)$.

3. Every logarithmic function is one-to-one.

4. Every logarithmic function has an inverse as a function.

5. If $b > 1$, the logarithmic function is increasing (and its graph is rising from left to right).

6. If $0 < b < 1$, the logarithmic function is decreasing (and its graph is falling from left to right).

7. The graph of every logarithmic function passes through the point $(1, 0)$ in the plane containing the graph.

8. The vertical axis of the plane containing the graph of a logarithmic function is a vertical asymptote for the graph.

9. If $b = e$, then $y = \log_e(x)$ is the **natural logarithmic function**. [Note: $\log_e(x) = \ln(x)$.]

In Exercises 28-34, sketch the graph of the given logarithmic function.

28. $y = \log_3(x)$

29. $y = \log_{1/3}(x)$

30. $y = \log_4(x) - 5$

31. $y = -\log_2(x)$

32. $y = \log_4(x) + 3$

33. $y = \ln(x)$

34. $y = \dfrac{\log_3(x)}{5}$

7.4 Differentiation and Integration of Logarithmic Functions

We have the following rules for differentiation of logarithmic functions: ($b > 0$, $b \neq 1$, u is a differentiable function of x)

Derivative of $\ln(x)$: $\dfrac{d}{dx}[\ln(x)] = \dfrac{1}{x}$, provided that $x > 0$.

Derivative of $\ln(u)$: $\dfrac{d}{dx}d[\ln(u)] = \dfrac{1}{u}\dfrac{du}{dx}$, provided that $u(x) > 0$.

Derivative of $\log_b(x)$: $\dfrac{d}{dx}[\log_b(x)] = \dfrac{1}{x\,\ln(b)}$, provided that $x > 0$.

Derivative of $\log_b(u)$: $\dfrac{d}{dx}[\log_b(u)] = \dfrac{1}{u\,\ln(b)}\dfrac{du}{dx}$, provided that $u(x) > 0$.

In Exercises 35-41, determine a simplified expression for $f'(x)$.

35. $f(x) = \log_3(x)$

36. $f(x) = \ln(3x)$

37. $f(x) = 10\ln(x^2 + 1)$

38. $f(x) = [2 + \ln(x)]^3$

39. $f(x) = x\ln(x)$

40. $f(x) = \dfrac{2x}{\ln(x)}$

41. $f(x) = (x - 3)\log_5(x)$

We have the following rules for integrals involving logarithmic functions: ($b > 0$, $b \neq 1$, u is a differentiable function of x)

Integral of $\dfrac{1}{x}$: $\displaystyle\int \dfrac{1}{x}dx = \ln|x| + C$, provided that $x \neq 0$.

Integral of $\dfrac{1}{u}$: $\displaystyle\int \dfrac{1}{u}u'(x)dx = \int \dfrac{1}{u}du = \ln|u| + C$, provided that $u(x) \neq 0$.

In Exercises 42-48, integrate.

42. $\int \dfrac{1}{2x}\, dx$

43. $\int \dfrac{x}{x^2 + 1}\, dx$

44. $\int \dfrac{4x^2 + 3x}{x^2}\, dx$

45. $\int \dfrac{e^x}{1 + e^x}\, dx$

46. $\int \dfrac{\ln(x)}{x}\, dx$

47. $\int \dfrac{x + 4}{x^2 + 8x + 1}\, dx$

48. $\int \dfrac{x}{x + 1}\, dx$

7.5 Exponential Growth and Decay

Definition: The equation $y = Ce^{kt}$ defines **exponential growth (or decay)** where C is the initial amount (or value) and k is the constant of proportionality.

An example of exponential growth is continuous compound interest. An example of exponential decay is radioactive decay.

---

49. If $10,000 is invested at an annual interest rate of 8% compounded continuously, how long will it take for the investment to double?

50. Repeat Exercise 49 if r = 11%.

51. If $8000 is invested for a period of 10 years and the interest is compounded continuously, at what rate must it be invested to double in value?

52. Repeat Exercise 51 if t = 150 months.

Definition: The time that it takes for a substance to decay to one-half of its current mass is called its **half-life**.

---

53. Radioactive radium has a half-life of approximately 1600 years. If we begin with 1000 grams of the radium, how much of it would remain after 2000 years?

54. The half-life of carbon 14, commonly used to determine the age of remains, is approximately 5570 years. If we begin with 75 grams of carbon 14, how much of it would remain after 9500 years?

55. A bone was uncovered at an archaeological dig and was found to have lost 75% of its carbon 14. How old is the bone? (Refer to Exercise 54.)

✓ Check Yourself

1. **(Exponential functions)**

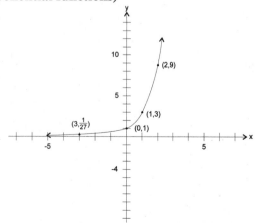

2. **(Exponential functions)**

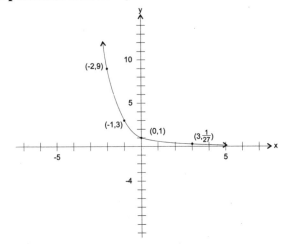

3. **(Exponential functions)**

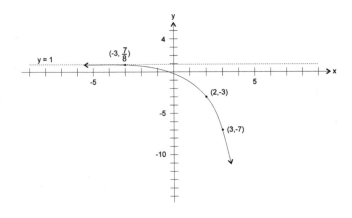

4. **(Exponential functions)**

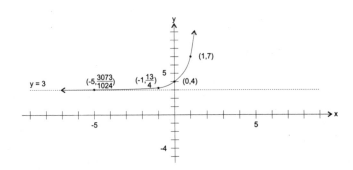

5. **(Exponential functions)**

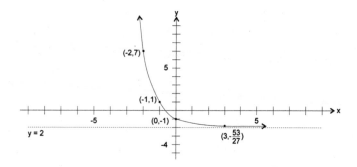

6. **(Exponential functions)**

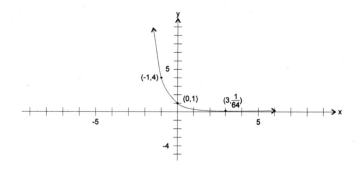

7. **(Exponential functions)**

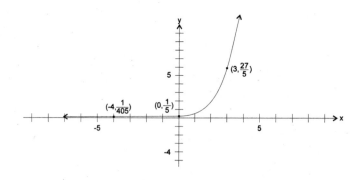

8. $A = p\left(1 + \dfrac{r}{n}\right)^{nt}$

    $= \$1000\left(1 + \dfrac{0.08}{2}\right)^{(2)(3)}$

    $= \$1000(1 + 0.04)^6$

    $= \$1000(1.04)^6$

    $\approx \$1265.32$ (**Compound interest**)

9. $A = p\left(1 + \dfrac{r}{n}\right)^{nt}$

    $= \$2000\left(1 + \dfrac{0.06}{12}\right)^{(12)(3.5)}$

    $= \$2000(1 + 0.005)^{42}$

    $= \$2000(1.005)^{42}$

    $\approx \$2466.07$ (**Compound interest**)

10. $A = p\left(1 + \dfrac{r}{n}\right)^{nt}$

    $= \$2000\left(1 + \dfrac{0.075}{12}\right)^{(12)(10)}$

    $= \$2000(1 + 0.00625)^{120}$

    $= \$2000(1.00625)^{120}$

    $\approx \$4224.13$ (**Compound interest**)

11. $A = Pe^{rt}$

    $= \$800e^{(0.07)(1)}$

    $= \$800e^{0.07}$

    $\approx \$858.01$ (**Continuous compound interest**)

12. $A = Pe^{rt}$

    $= \$900e^{(0.08)(2.5)}$

    $= \$900e^{0.2}$

    $\approx \$1099.26$ (**Continuous compound interest**)

13. $A = Pe^{rt}$

    $= \$20,000e^{(0.1)(0.5)}$

    $= \$20,000e^{0.05}$

    $\approx \$21,025.42$ (**Continuous compound interest**)

14. $f(x) = 3^x$ for all real values of x.

    $f'(x) = \dfrac{d}{dx}(3^x) = 3^x \ln(3)$ for all real values of x. (**Derivative of b^x**)

15. $f(x) = e^{3x}$ for all real values of x.

$f'(x) = \dfrac{d}{dx}(e^{3x}) = e^{3x}\dfrac{d}{dx}(3x) = 3e^{3x}$ for all real values of x. (**Derivative of e^u**)

16. $f(x) = 50e^{x^2+1}$ for all real values of x.

$f'(x) = \dfrac{d}{dx}(50e^{x^2+1}) = 50e^{x^2+1}\dfrac{d}{dx}(x^2 + 1) = 50e^{x^2+1}(2x) = 100xe^{x^2+1}$ for all real values of x.
(**Derivative of ce^u**)

17. $f(x) = (2 + e^x)^3$ for all real values of x.

$f'(x) = \dfrac{d}{dx}(2 + e^x)^3 = 3(2 + e^x)^2\dfrac{d}{dx}(2 + e^x) = 3(2 + e^x)^2(e^x) = 3e^x(2 + e^x)^2$ for all real values of x.
(**Derivative involving e^x**)

18. $f(x) = xe^x$ for all real values of x.

$f'(x) = \dfrac{d}{dx}(xe^x) = x\dfrac{d}{dx}(e^x) + e^x\dfrac{d}{dx}(x) = xe^x + e^x(1) = xe^x + e^x = e^x(x + 1)$ for all real values of x.
(**Derivative involving e^x**)

19. $f(x) = \dfrac{2x}{e^x}$ for all real values of x.

$f'(x) = \left(\dfrac{2x}{e^x}\right)' = \dfrac{e^x\dfrac{d}{dx}(2x) - 2x\dfrac{d}{dx}(e^x)}{(e^x)^2}$

$= \dfrac{e^x(2) - 2xe^x}{e^{2x}} = \dfrac{2e^x - 2xe^x}{e^{2x}} = \dfrac{2e^x(1 - x)}{e^{2x}} = \dfrac{2(1 - x)}{e^x}$ for all real values of x. (**Derivative involving e^x**)

20. $f(x) = (3 - x)(5^x)$ for all real values of x.

$f'(x) = \dfrac{d}{dx}[(3 - x)(5^x)] = (3 - x)\dfrac{d}{dx}(5^x) + 5^x\dfrac{d}{dx}(3 - x)$

$= (3 - x)(5^x)[\ln(5)] + 5^x(-1) = (3 - x)[\ln(5)](5^x) - 5^x$

$= 5^x[(3 - x)\ln(5) - 1]$ for all real values of x. (**Derivative involving b^x**)

21. $\int e^{2x}\, dx$ for all real values of x.

Let $u = 2x$. Then, $du = \dfrac{d}{dx}(2x)\, dx = 2\, dx$.

Substituting, we have

$\int e^{2x}\, dx = \int e^u\left(\dfrac{1}{2}\right)du = \dfrac{1}{2}\int e^u\, du = \dfrac{1}{2}e^u + C.$

Substituting 2x for u, we have

$\int e^{2x}\, dx = \dfrac{1}{2}e^u + C = \dfrac{1}{2}e^{2x} + C$ for all real values of x. (**Integral of e^u**)

22. $\int xe^{2x^2} dx$ for all real values of x.

Let $u = 2x^2$. Then, $du = \dfrac{d}{dx}(2x^2)\,dx = 4x\,dx$.

Substituting, we have

$$\int xe^{2x^2}\,dx = \int e^{2x^2}(x\,dx) = \int e^u\left(\frac{1}{4}\right)du = \frac{1}{4}\int e^u\,du = \frac{1}{4}e^u + C.$$

Substituting $2x^2$ for u, we have

$$\int xe^{2x^2}\,dx = \frac{1}{4}e^u + C = \frac{1}{4}e^{2x^2} + C \text{ for all real values of x. } (\textbf{Integral of } e^u)$$

23. $\displaystyle\int \frac{5^{2/x}}{x^2}\,dx$ for all real values of $x \neq 0$.

Let $u = \dfrac{2}{x}$. Then, $du = \dfrac{d}{dx}\left(\dfrac{2}{x}\right)dx = \dfrac{-2}{x^2}\,dx$

Substituting, we have

$$\int \frac{5^{2/x}}{x^2}\,dx = \int 5^{2/x}\left(\frac{1}{x^2}\,dx\right) = \frac{-1}{2}\int 5^{2/x}\left(\frac{-2}{x^2}\right)dx = \frac{-1}{2}\int 5^u\,du = \frac{-5^u}{2\ln(5)} + C$$

Substituting $\dfrac{2}{x}$ for u, we have

$$\int \frac{5^{2/x}}{x^2}\,dx = \frac{-5^u}{2\ln(5)} + C = \frac{-5^{2/x}}{2\ln(5)} + C \text{ for all real values of x. } (\textbf{Integral of } b^u)$$

24. $\int (e^x + e^{-x})^2\,dx$ for all real values of x.

$$\int (e^x + e^{-x})^2\,dx = \int (e^{2x} + 2e^x e^{-x} + e^{-2x})\,dx$$

$$= \int (e^{2x} + 2 + e^{-2x})\,dx = \int e^{2x}\,dx + 2\int dx + \int e^{-2x}\,dx$$

Let $u = 2x$. Then $du = \dfrac{d}{dx}(2x)\,dx = 2\,dx$.

Let $v = -2x$. Then $dv = \dfrac{d}{dx}(-2x)\,dx = -2\,dx$.

Substituting, we have

$$\int e^{2x}\,dx + 2\int dx + \int e^{-2x}\,dx = \int e^u\left(\frac{1}{2}\right)du + 2x + \int e^v\left(\frac{-1}{2}\right)dv$$

$$= \frac{1}{2}\int e^u\,du + 2x - \frac{1}{2}\int e^v\,dv = \frac{1}{2}e^u + 2x - \frac{1}{2}e^v + C.$$

Substituting 2x for u and $-2x$ for v, we have

$$\int (e^x + e^{-x})^2\,dx = \frac{1}{2}e^u + 2x - \frac{1}{2}e^v + C = \frac{1}{2}e^{2x} + 2x - \frac{1}{2}e^{-2x} + C \text{ for all real values of x. } (\textbf{Integral}$$

$\textbf{involving } e^u)$

25. $\int e^x(1 - e^x)^4 \, dx$ for all real values of x.

Let $u = 1 - e^x$. Then

$$du = \frac{d}{dx}(1 - e^x) \, dx = -e^x \, dx$$

Substituting, we have

$$\int e^x(1 - e^x)^4 \, dx = \int (1 - e^x)^4 (e^x \, dx) = \int u^4 \, (-du) = -\int u^4 \, du = \frac{-u^5}{5} + C.$$

Substituting $1 - e^x$ for u, we have

$$\int e^x(1 - e^x)^4 \, dx = \frac{-u^5}{5} + C = \frac{-(1 - e^x)^5}{5} + C \text{ for all real values of x. } \textbf{(Integral involving } e^x\textbf{)}$$

26. $\int (4^{3x-1} + 2) \, dx$ for all real values of x.

$$\int (4^{3x-1} + 2) \, dx = \int 4^{3x-1} \, dx + 2 \int dx = \int 4^{3x-1} \, dx + 2x + C.$$

Let $u = 3x - 1$. Then,

$$du = \frac{d}{dx}(3x - 1) \, dx = 3 \, dx \, .$$

Substituting, we have

$$\int (4^{3x-1} + 2) \, dx = \int 4^u \left(\frac{1}{3} du\right) + 2x = \frac{1}{3}\int 4^u \, du + 2x = \left(\frac{1}{3}\right)\frac{4^u}{\ln(4)} + 2x + C.$$

Substituting $3x - 1$ for u, we have

$$\int (4^{3x-1} + 2) \, dx = \frac{4^u}{3 \ln(4)} + 2x + C = \frac{4^{3x-1}}{3 \ln(4)} + 2x + C \text{ for all real values of x. } \textbf{(Integral involving } b^u\textbf{)}$$

27. $\int (e^{3x} + e^{5x}) \, dx$ for all real values of x.

$$\int (e^{3x} + e^{5x}) \, dx = \int e^{3x} \, dx + \int e^{5x} \, dx$$

Let $u = 3x$. Then, $du = \frac{d}{dx}(3x) \, dx = 3 \, dx.$

Let $v = 5x$. Then, $dv = \frac{d}{dx}(5x) \, dx = 5 \, dx.$

Substituting, we have

$$\int (e^{3x} + e^{5x}) \, dx = \int e^u \left(\frac{1}{3}du\right) + \int e^v \left(\frac{1}{5}dv\right)$$

$$= \frac{1}{3}\int e^u \, du + \frac{1}{5}\int e^v \, dv = \frac{1}{3}e^u + \frac{1}{5}e^v + C.$$

Substituting 3x for u and 5x for v, we have

$$\int (e^{3x} + e^{5x}) \, dx = \frac{1}{3}e^u + \frac{1}{5}e^v + C = \frac{1}{3}e^{3x} + \frac{1}{5}e^{5x} + C \text{ for all real values of x. } \textbf{(Integral involving } e^u\textbf{)}$$

28. **(Logarithmic functions)**

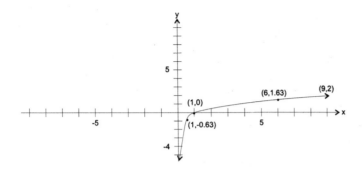

29. **(Logarithmic functions)**

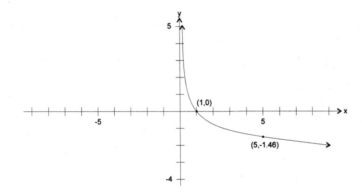

30. **(Logarithmic functions)**

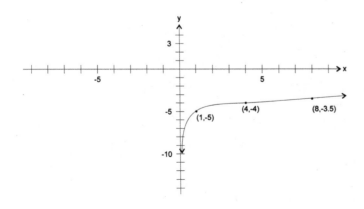

31. **(Logarithmic functions)**

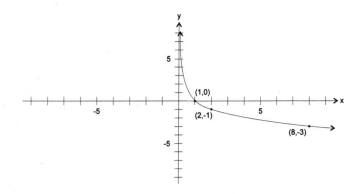

32. **(Logarithmic functions)**

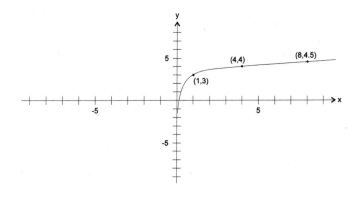

33. **(Logarithmic functions)**

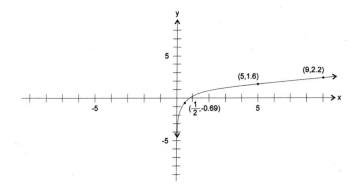

34. **(Logarithmic functions)**

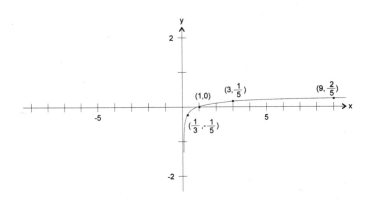

35. $f(x) = \log_3(x)$ for all real values of $x > 0$.

$$f'(x) = \frac{d}{dx}[\log_3(x)] = \frac{1}{x \ln(3)} \text{ for all real values of } x > 0. \text{ (\textbf{Derivative of } } \log_b(x))$$

36. $f(x) = \ln(3x)$ for all real values of $x > 0$.

$$f'(x) = \frac{d}{dx}[\ln(3x)] = \frac{1}{3x}\frac{d}{dx}(3x) = \frac{1}{3x}(3) = \frac{1}{x} \text{ for all real values of } x > 0. \text{ (\textbf{Derivative of } } \ln(u))$$

37. $f(x) = 10 \ln(x^2 + 1)$ for all real values of x.

$$f'(x) = \frac{d}{dx}[10 \ln(x^2 + 1)] = 10\frac{d}{dx}[\ln(x^2 + 1)]$$

$$= \left(\frac{10}{x^2 + 1}\right)\frac{d}{dx}(x^2 + 1) = \frac{10}{x^2 + 1}(2x) = \frac{20x}{x^2 + 1} \text{ for all real values of } x. \text{ (\textbf{Derivative involving } } \ln(u))$$

38. $f(x) = [2 + \ln(x)]^3$ for all real values of $x > 0$.

$$f'(x) = \frac{d}{dx}[2 + \ln(x)]^3 = 3[2 + \ln(x)]^2\frac{d}{dx}[2 + \ln(x)]$$

$$= 3[2 + \ln(x)]^2\left(\frac{1}{x}\right) = \left(\frac{3}{x}\right)[2 + \ln(x)]^2 \text{ for all real values of } x > 0. \text{ (\textbf{Derivative involving } } \ln(x))$$

39. $f(x) = x \ln(x)$ for all real values of $x > 0$.

$$f'(x) = \frac{d}{dx}[x \ln(x)] = x\frac{d}{dx}[\ln(x)] + \ln(x)\frac{d}{dx}(x)$$

$$= x\left(\frac{1}{x}\right) + \ln(x)(1) = 1 + \ln(x) \text{ for all real values of } x > 0. \text{ (\textbf{Derivative involving } } \ln(x))$$

40. $f(x) = \frac{2x}{\ln(x)}$ for all real values of $x > 0$ and such that $x \neq 1$.

$$f'(x) = \frac{d}{dx}\left(\frac{2x}{\ln(x)}\right) = \frac{\ln(x)\frac{d}{dx}(2x) - 2x\frac{d}{dx}[\ln(x)]}{[\ln(x)]^2}$$

$$= \frac{\ln(x)\,(2) - (2x)\left(\dfrac{1}{x}\right)}{[\ln(x)]^2} = \frac{2\ln(x) - 2}{[\ln(x)]^2} \text{ for all real values of } x > 0 \text{ and such that } x \ne 1.$$

(Derivative involving ln(x))

41. $f(x) = (x - 3)\log_5(x)$ for all real values of $x > 0$.

$$f'(x) = \frac{d}{dx}\left[(x - 3)\log_5(x)\right]$$

$$= (x - 3)\frac{d}{dx}[\log_5(x)] + \log_5(x)\frac{d}{dx}(x - 3)$$

$$= (x - 3)\left(\frac{1}{x\,\ln(5)}\right) + \log_5(x)\,(1)$$

$$= \frac{x - 3}{x\,\ln(5)} + \log_5(x) \text{ for all real values of } x > 0. \textbf{ (Derivative involving } \log_b(x)\textbf{)}$$

42. $\displaystyle\int \frac{1}{2x}\,dx$ for all real values of $x \ne 0$.

$$\int \frac{1}{2x}\,dx = \frac{1}{2}\int \frac{1}{x}\,dx = \frac{1}{2}\ln|\,x\,| + C \text{ for all real values of } x \ne 0. \textbf{ (Integral involving } \tfrac{1}{x}\textbf{)}$$

43. $\displaystyle\int \frac{x}{x^2 + 1}\,dx$ for all real values of x.

Let $u = x^2 + 1$. Then,

$$du = \frac{d}{dx}(x^2 + 1)\,dx = 2x\,dx.$$

Substituting, we have

$$\int \frac{x}{x^2 + 1}\,dx = \int \frac{1}{u}\left(\frac{1}{2}du\right) = \frac{1}{2}\int \frac{1}{u}\,du = \frac{1}{2}\ln|\,u\,| + C.$$

Substituting $x^2 + 1$ for u, we have

$$\int \frac{x}{x^2 + 1}\,dx = \frac{1}{2}\ln|\,u\,| + C = \frac{1}{2}\ln(x^2 + 1) + C \text{ for all real values of x.} \textbf{ (Integral involving } \tfrac{1}{u}\textbf{)}$$

44. $\displaystyle\int \frac{4x^2 + 3x}{x^2}\,dx$ for all real values of $x \ne 0$.

$$\int \frac{4x^2 + 3x}{x^2}\,dx = \int \left(4 + \frac{3}{x}\right)dx = 4\!\int dx + 3\int \frac{1}{x}\,dx$$

$$= 4x + 3\ln|\,x\,| + C \text{ for all real values of } x \ne 0. \textbf{ (Integral involving } \tfrac{1}{x}\textbf{)}$$

45. $\displaystyle\int \frac{e^x}{1 + e^x}\,dx$ for all real values of x.

Let $u = 1 + e^x$. Then,

$$du = \frac{d}{dx}(1 + e^x)\,dx = e^x\,dx.$$

Substituting, we have

$$\int \frac{e^x}{1+e^x}\,dx = \int \frac{1}{1+e^x}(e^x\,dx) = \int \frac{1}{u}\,du = \ln|\,u\,| + C.$$

Substituting $1 + e^x$ for u, we have

$$\int \frac{e^x}{1+e^x}\,dx = \ln|\,u\,| + C = \ln(1 + e^x) + C \text{ for all real values of x. (\textbf{Integral of } $\frac{1}{u}$)}$$

46. $\displaystyle\int \frac{\ln(x)}{x}\,dx$ for all real values of $x > 0$.

Let $u = \ln(x)$. Then,

$$du = \frac{d}{dx}[\ln(x))]\,dx = \frac{1}{x}\,dx.$$

Substituting, we have

$$\int \frac{\ln(x)}{x}\,dx = \int \ln(x)\left(\frac{1}{x}dx\right) = \int u\,du = \frac{u^2}{2} + C$$

Substituting ln(x) for u, we have

$$\int \frac{\ln(x)}{x}\,dx = \frac{u^2}{2} + C = \frac{[\ln(x)]^2}{2} + C \text{ for all real values of } x > 0. \textbf{ (Integral involving The General}$$
Power Rule)

47. $\displaystyle\int \frac{x+4}{x^2+8x+1}\,dx$ for all real values of $x \neq -4 \pm \sqrt{15}$.

Let $u = x^2 + 8x + 1$. Then,

$$du = \frac{d}{dx}(x^2 + 8x + 1)\,dx = (2x + 8)\,dx$$

$$= 2(x + 4)\,dx.$$

Substituting, we have

$$\int \frac{x+4}{x^2+8x+1}\,dx = \int \frac{1}{x^2+8x+1}(x+4)\,dx = \int \frac{1}{u}\left(\frac{1}{2}du\right) = \frac{1}{2}\int \frac{1}{u}du = \frac{1}{2}\ln|\,u\,| + C$$

Substituting $x^2 + 8x + 1$ for u, we have

$$\int \frac{x+4}{x^2+8x+1}\,dx = \frac{1}{2}\ln|\,u\,| + C = \frac{1}{2}\ln(x^2 + 8x + 1) + C \text{ for all real values of } x \neq -4 \pm \sqrt{15}. \textbf{ (Integral}$$
involving $\frac{1}{u}$)

48. $\displaystyle\int \frac{x}{x+1}\,dx$ for all real values of $x \neq -1$.

$$\int \frac{x}{x+1}\,dx = \int (1 - \frac{1}{x+1})\,dx = \int dx - \int \frac{1}{x+1}\,dx = x - \int \frac{1}{x+1}\,dx$$

Let $u = x + 1$. Then,

$$du = \frac{d}{dx}(x + 1)\,dx = (1)\,dx = dx.$$

Substituting, we have

$$\int \frac{x}{x+1}\,dx = x - \int \frac{1}{u}\,du = x - \ln|u| + C.$$

Substituting x + 1 for u, we have

$$\int \frac{x}{x+1}\,dx = x - \ln|u| + C = x - \ln|x+1| + C \text{ for all real values of } x \ne -1. \textbf{ (Integral involving } \frac{1}{u}\textbf{)}$$

49. $A = Pe^{rt}$

$\$20,000 = \$10,000e^{0.08t}$

$2 = e^{0.08t}$

$\ln(2) = \ln(e^{0.08t})$

$\ln(2) = (0.08t)\ln(e)$

$\ln(2) = (0.08t)(1)$

$\ln(2) = 0.08t$

$t = \dfrac{\ln(2)}{0.08}$

$t \approx 8.7$ yr **(Exponential growth)**

50. $A = Pe^{rt}$

$\$20,000 = \$10,000e^{0.11t}$

$2 = e^{0.11t}$

$\ln(2) = \ln(e^{0.11t})$

$\ln(2) = (0.11t)\ln(e)$

$\ln(2) = (0.11t)(1)$

$\ln(2) = 0.11t$

$t = \dfrac{\ln(2)}{0.11}$

$t \approx 6.3$ yr **(Exponential growth)**

51. $A = Pe^{rt}$

$\$16,000 = \$8000e^{10r}$

$2 = e^{10r}$

$\ln(2) = \ln(e^{10r})$

$\ln(2) = (10r)\ln(e)$

$\ln(2) = (10r)(1)$

$\ln(2) = 10r$

$r = \dfrac{\ln(2)}{10}$

$t \approx 6.93\%$ **(Exponential growth)**

52. $A = Pe^{rt}$

$$\$16{,}000 = \$8000e^{12.5r}$$

$$2 = e^{12.5r}$$

$$\ln(2) = \ln(e^{12.5r})$$

$$\ln(2) = (12.5r)\ln(e)$$

$$\ln(2) = (12.5r)(1)$$

$$\ln(2) = 12.5r$$

$$r = \frac{\ln(2)}{12.5}$$

$$t \approx 5.55\% \textbf{ (Exponential growth)}$$

53. $y = Ce^{kt}$

When t = 0 (years), y = 1000 (grams). Hence,

$$1000 = Ce^0 = C(1) = C.$$

When t = 1600 (years), $y = \frac{1}{2}(1000) = 500$ (grams). We have

$$500 = 1000e^{1600k}$$

$$0.5 = e^{1600k}$$

$$\ln(0.5) = 1600\,k$$

$$k = \frac{\ln(0.5)}{1600}$$

$$k \approx -.000433$$

Hence, y as a function of t is given by

$$y \approx 1000e^{-0.000433t}$$

When t = 2000 (years), we have

$$y \approx 1000e^{(-0.000433)(2000)}$$

$$y \approx 1000(0.42063)$$

$$y \approx 420.63 \text{ (grams)}$$

Therefore, approximately 421 grams of the radium would still be present after 2000 years. **(Exponential decay)**

54. $y = Ce^{kt}$

When t = 0 (years), y = 75 (grams). Hence,

$$75 = Ce^0 = C(1) = C.$$

When t = 5570 (years), $y = \frac{1}{2}(75) = 37.5$ (grams). We have

$$37.5 = 75e^{5570k}$$

$$0.5 = e^{5570k}$$

$$\ln(0.5) = 5570 \, k$$

$$k = \frac{\ln(0.5)}{5570}$$

$$k \approx -.0001244$$

Hence, y as a function of t is given by

$$y \approx 75e^{-0.0001244t}$$

When t = 9500 (years), we have

$$y \approx 75e^{(-0.0001244)(9500)}$$

$$y \approx 75(0.3067)$$

$$y \approx 23 \text{ (grams)}$$

Therefore, approximately 23 grams of the carbon 14 would still be present after 9500 years. (**Exponential decay**)

55. Since 75% of the carbon 14 has been lost, then 25% still remains. Let y_0 = amount of carbon 14 (in grams) when t = 0.

Then,

$$y = y_0 e^{kt}.$$

In Exercise 54, we determined that k $\approx$ −0.0001244. Hence,

$$y \approx y_0 e^{-0.0001244t}$$

When $y = 0.25y_0$, we have

$$0.25y_0 \approx y_0 e^{-0.0001244t}$$

$$0.25 \approx e^{-0.0001244t}$$

$$\ln(0.25) \approx -0.0001244t$$

$$t \approx \frac{\ln(0.25)}{-0.0001244}$$

$$t \approx 11{,}143.85$$

Therefore, the bone is approximately 11,143.85 years old. (**Exponential decay**)

Grade Yourself

Circle the numbers of the questions you missed, then fill in the total incorrect for each topic. If you answered more than three questions incorrectly, you need to focus on that topic. (If a topic has less than three questions and you had at least one wrong, we suggest you study that topic also. Read your textbook, a review book, or ask your teacher for help.)

Subject: Exponential and Logarithmic Functions

Topic	Question Numbers	Number Incorrect
Exponential functions	1, 2, 3, 4, 5, 6, 7	
Compound interest	8, 9, 10, 11, 12, 13	
Derivative of b^x	14, 20	
Derivative of e^u	15	
Derivative of ce^u	16	
Derivative of e^x	17, 18, 19	
Integral of e^u	21, 22, 24	
Integral involving b^u	23, 26	
Integral involving e^x	25	
Integral involving e^u	27	
Logarithmic functions	28, 29, 30, 31, 32, 33, 34	
Derivative of $\log_b(x)$	35, 41	
Derivative of $\ln(u)$	36, 37	
Derivative of $\ln(x)$	38, 39, 40	
Integral involving $\frac{1}{x}$	42, 44	
Integral involving $\frac{1}{u}$	43, 47, 48	
Integral of $\frac{1}{u}$	45	
Integral involving The General Power Rule	46	
Exponential growth	49, 50, 51, 52	
Exponential decay	53, 54, 55	

Techniques of Integration

8

 Test Yourself

8.1 Integration by Substitution

To integrate by substitution, we let u be some function of the given independent variable, say x. This function of x is usually part of the integrand. We then solve for x and dx in terms of u and du. With the substitution complete, we try to integrate using formulas already studied. If we cannot integrate, we often can try a different substitution. If the integration can be done, the result will be a function of u. Substitute back so that the final result is a function of x.

In Exercises 1-11, use a substitution to determine the indefinite integral.

1. $\int \dfrac{x}{(x+1)^4}\,dx$

2. $\int t^2\sqrt{t+1}\,dt$

3. $\int \dfrac{1}{\sqrt{1-x}}\,dx$

4. $\int x^2(x^3-2)^7\,dx$

5. $\int \dfrac{3p-1}{\sqrt{p+1}}\,dp$

6. $\int \dfrac{y}{(y-2)^2}\,dy$

7. $\int \dfrac{t+1}{t^2+2t}\,dt$

8. $\int xe^{x^2}\,dx$

9. $\int \dfrac{e^y}{1+e^y}\,dy$

10. $\int \dfrac{2e^{-1/p}}{p^2}\,dp$

11. $\int \dfrac{1}{t\,\ln(t)}\,dt$

8.2 Integration by Parts

Another basic technique of integration is known as *integration by parts*. Since the method is based upon the differentiation formula for a product of two functions, it will be especially useful for integrating products.

Integration by Parts: $\int u\,dv = uv - \int v\,du$

Observe that this formula expresses a given integral, $\int u\,dv$, in terms of another integral, $\int v\,du$. This new integral, $\int v\,du$, may be easier to integrate than the original integral.

In Exercise 12-23, integrate using the method of integration by parts.

12. $\int \ln(x)\,dx$

13. $\int x\,\ln(x)\,dx$

14. $\int ye^{3y}\, dy$

15. $\int \dfrac{\ln(t)}{t}\, dt$

16. $\int te^{-t}\, dt$

17. $\int \dfrac{p}{e^{4p}}\, dp$

18. $\int \ln(p^2)\, dp$

19. $\int x^2 e^x\, dx$

20. $\int x\, \ln(1+x)\, dx$

21. $\int y^3\, \ln(y)\, dy$

22. $\int [\ln(x)]^2\, dx$

23. $\int x^2 e^{-x}\, dx$

8.3 Average Value of a Function

In Chapter 6, we used definite integrals to determine the area of certain plane regions. Definite integrals can also be used to determine the average value of a function over an interval.

Definition: Let f be a continuous function on the interval [a, b]. Then, *the average value of the function* f on the interval [a, b] is given by

$$\text{Avg. value of } f \text{ on } [a,\, b] = \frac{1}{b-a}\int_a^b f(x)\, dx$$

In Exercises 24–30, determine the average value of the given function on the indicated interval.

24. $f(x) = x;\ [0,\, 1]$

25. $f(x) = x^2 + 1;\ [-1,\, 2]$

26. $g(x) = x^3 + 2;\ [-2,\, 2]$

27. $h(x) = x\sqrt{4 - x^2}\ ;\ [0,\, 2]$

28. $p(x) = \dfrac{x^2 + 3}{x^2}\ ;\ [1,\, 4]$

29. $t(x) = \sqrt{x}\,(x - 1);\ [0,\, 1]$

30. $t(x) = x^3 - 2x;\ [-4,\, -1]$

✓ Check Yourself

1. $\int \dfrac{x}{(x+1)^4}\, dx$ for all real values of $x \neq -1$.

Let $u = x + 1$. Then,

$x = u - 1$ and

$dx = \dfrac{d}{du}(u - 1)\, du = (1)\, du = du.$

Substituting, we have

$$\int \frac{x}{(x+1)^4}\, dx = \int \frac{u-1}{u^4}\, du = \int \left(\frac{1}{u^3} - \frac{1}{u^4}\right) du = \int (u^{-3} - u^{-4})\, du$$

$$= \left[\frac{u^{-2}}{-2} - \frac{u^{-3}}{-3}\right] + C = \frac{-1}{2u^2} + \frac{1}{3u^3} + C.$$

Substituting $x + 1$ for u, we have

$$\int \frac{x}{(x+1)^4}\, dx = \frac{-1}{2(x+1)^2} + \frac{1}{3(x+1)^3} + C = \frac{-3(x+1)+2}{6(x+1)^3} + C$$

$$= \frac{-3x-1}{6(x+1)^3} + C \text{ for all real values of } x \neq -1. \textbf{ (Integration by substitution)}$$

2. $\int t^2\sqrt{t+1}\ dt$ for all real values of $t \geq -1$.

    Let $u = t + 1$. Then,

    $u^2 = t + 1$. Hence,

    $t = u^2 - 1$ and

    $dt = \dfrac{d}{du}(u^2 - 1)\ du = 2u\ du.$

    Substituting, we have

    $\int t^2\sqrt{t+1}\ dt = \int (u^2 - 1)^2 u(2u)\ du = 2\int u^2(u^2 - 1)^2\ du$

    $= 2\int u^2(u^4 - 2u^2 + 1)\ du = 2\int (u^6 - 2u^4 + u^2)\ du$

    $= 2\left[\dfrac{u^7}{7} - \dfrac{2u^5}{5} + \dfrac{u^3}{3}\right] + C$

    Substituting $(t + 1)^{1/2}$ for u, we have

    $\int t^2\sqrt{t+1}\ dt = \dfrac{2}{7}(t+1)^{7/2} - \dfrac{4}{5}(t+1)^{5/2} + \dfrac{2}{3}(t+1)^{3/2} + C$ for all real values of $t \geq -1$. **(Integration by substitution)**

3. $\int \dfrac{1}{\sqrt{1-x}}\ dx$ for all $x < 1$.

    Let $u = \sqrt{1-x}$. Then,

    $u^2 = 1 - x$. Hence,

    $x = 1 - u^2$ and

    $dx = \dfrac{d}{du}(1 - u^2)\ du = -2u\ du.$

    Substituting, we have

    $\int \dfrac{1}{\sqrt{1-x}}\ dx = \int \dfrac{1}{u}(-2u)du = -2\int du = -2u + C.$

    Substituting $\sqrt{1-x}$ for u, we have

    $\int \dfrac{1}{\sqrt{1-x}}\ dx = -2\sqrt{1-x} + C$ for all real values of $x < 1$. **(Integration by substitution)**

4. $\int x^2(x^3 - 2)^7\ dx$ for all real values of x.

    Let $u = x^3 - 2$. Then, $du = \dfrac{d}{dx}(x^3 - 2)\ dx = 3x^2\ dx$. Hence,

    $x^2\ dx = \dfrac{1}{3}\ du$. We have

    $\int x^2(x^3 - 2)^7\ dx = \int (x^3 - 2)^7 x^2\ dx = \int u^7\left(\dfrac{1}{3}\right)du = \dfrac{1}{3}\int u^7\ du = \dfrac{1}{3} \cdot \dfrac{u^8}{8} + C = \dfrac{u^8}{24} + C$

    Substituting $x^3 - 2$ for u, we have

$\int x^2(x^3-2)^7 dx = \dfrac{1}{24}(x^3-2)^8 + C$ for all real values of x. **(Integration by substitution)**

5. $\int \dfrac{3p-1}{\sqrt{p+1}} dp$ for all real values of p > −1.

Let $u = \sqrt{p+1}$. Then,

$u^2 = p + 1$. Hence,

$p = u^2 - 1$ and

$dp = \dfrac{d}{du}(u^2-1)\,du = 2u\,du.$

Substituting, we have

$\int \dfrac{3p-1}{\sqrt{p+1}} dp = \int \dfrac{3(u^2-1)-1}{u}(2u)du = 2\int[3(u^2-1)-1]\,du$

$2\int(3u^2-4)\,du = 2(u^3-4u) + C = 2u^3 - 8u + C.$

Substituting $\sqrt{p+1}$ for u, we have

$\int \dfrac{3p-1}{\sqrt{p+1}} dp = 2(p+1)^{3/2} - 8\sqrt{p+1} + C$ for all real values of p > −1. **(Integration by substitution)**

6. $\int \dfrac{y}{(y-2)^2} dy$ for all real values of y ≠ 2.

Let $u = y - 2$. Then,

$y = u + 2$ and

$dy = \dfrac{d}{du}(u+2)\,du = (1)\,du = du.$

Substituting, we have

$\int \dfrac{y}{(y-2)^2} dy = \int \dfrac{u+2}{u^2} du = \int\left(\dfrac{1}{u}+\dfrac{2}{u^2}\right)du = \ln|u| + 2\int u^{-2}\,du$

$= \ln|u| + \dfrac{2u^{-1}}{-1} + C = \ln|u| - \dfrac{2}{u} + C. \ (u \neq 0)$

Substituting y − 2 for u, we have

$\int \dfrac{y}{(y-2)^2} dy = \ln|y-2| - \dfrac{2}{y-2} + C$ for all real values of y ≠ 2. **(Integration by substitution)**

7. $\int \dfrac{t+1}{t^2+2t} dt$ for all real values of t ≠ −2, 0.

Let $u = t^2 + 2t$. Then,

$du = \dfrac{d}{dt}(t^2+2t)\,dt = (2t+2)\,dt = 2(t+1)\,dt.$ Hence,

$(t+1)\,dt = \dfrac{1}{2}\,du.$

Substituting, we have

$$\int \frac{t+1}{t^2+2t}\,dt = \int \frac{(t+1)\,dt}{t^2+2t} = \int \frac{(1/2)\,du}{u} = \frac{1}{2}\int \frac{1}{u}\,du = \frac{1}{2}\,\ln|\,u\,|+C. \quad (u \neq 0)$$

Substituting $t^2 + 2t$ for u, we have

$$\int \frac{t+1}{t^2+2t}\,dt = \frac{1}{2}\,\ln|\,t^2+2t\,|+C \text{ for all real values of } t \neq -2, 0. \textbf{ (Integration by substitution)}$$

8. $\int xe^{x^2}\,dx$ for all real values of x.

Let $u = x^2$. Then,

$$du = \frac{d}{dx}\,(x^2) = 2x\,dx. \text{ Hence,}$$

$$x\,dx = \frac{1}{2}\,du.$$

Substituting, we have

$$\int xe^{x^2}\,dx = \int e^{x^2}\,(x)\,dx = \int e^u\left(\frac{1}{2}\right)du = \frac{1}{2}\int e^u\,du = \frac{1}{2}\,e^u + C.$$

Substituting x^2 for u, we have

$$\int xe^{x^2}\,dx = \frac{1}{2}\,e^{x^2} + C \text{ for all real values of x. } \textbf{(Integration by substitution)}$$

9. $\int \frac{e^y}{1+e^y}\,dy$ for all real values of y.

Let $u = 1 + e^y$. Then,

$$du = \frac{d}{dy}\,(1 + e^y)\,dy = e^y\,dy.$$

Substituting, we have

$$\int \frac{e^y}{1+e^y}\,dy = \int \frac{e^y\,dy}{1+e^y} = \int \frac{du}{u} = \int \frac{1}{u}\,du = \ln|\,u\,|+C. \quad (u \neq 0)$$

Substituting $1 + e^y$ for u, we have

$$\int \frac{e^y}{1+e^y}\,dy = \ln(1 + e^y) + C \text{ for all real values of y. } \textbf{(Integration by substitution)}$$

10. $\int \frac{2e^{-1/p}}{p^2}\,dp$ for all real values of $p \neq 0$.

Let $u = e^{-1/p}$. Then,

$$du = \frac{d}{dp}\,(e^{-1/p})\,dp = e^{-1/p}\left(\frac{1}{p^2}\right)dp.$$

Substituting, we have

$$\int \frac{2e^{-1/p}}{p^2}\,dp = 2\int e^{-1/p}\left(\frac{1}{p^2}\right)dp = 2\int du = 2u + C.$$

Substituting $e^{-1/p}$ for u, we have

$$\int \frac{2e^{-1/p}}{p^2} \, dp = 2e^{-1/p} + C \text{ for all real values of p.} \quad \textbf{(Integration by substitution)}$$

11. $\int \frac{1}{t \, \ln(t)} \, dt$ for all real values of $t > 0$ and such that $t \neq 1$.

Let $u = \ln(t)$. Then,

$$du = \frac{d}{dt} (\ln(t)) \, dt = \frac{1}{t} \, dt.$$

Substituting, we have

$$\int \frac{1}{t \, \ln(t)} \, dt = \int \frac{1}{\ln(t)} \left(\frac{1}{t} \right) dt = \int \frac{1}{u} du = \ln|u| + C. \, (u \neq 0)$$

Substituting $\ln(t)$ for u, we have

$$\int \frac{1}{t \, \ln(t)} = \ln|\ln(t)| + C \text{ for all real values of } t > 0 \text{ and such that } t \neq 1. \, \textbf{(Integration by substitution)}$$

12. $\int \ln(x) \, dx$ for all real values of $x > 0$.

Let $u = \ln(x)$ and $dv = dx$. Then,

$$du = \frac{1}{x} \, dx \text{ and } v = x. \text{ We have}$$

$$\int \ln(x) \, dx = x \, \ln(x) - \int x \left(\frac{1}{x} \right) dx$$

$$= x \, \ln(x) - \int dx$$

$$= x \, \ln(x) - x + C \text{ for all real values of } x > 0. \quad \textbf{(Integration by parts)}$$

13. $\int x \, \ln(x) \, dx$ for all real values of $x > 0$.

Let $u = \ln(x)$ and $dv = x \, dx$. Then,

$$du = \frac{1}{x} \, dx \text{ and } v = \frac{x^2}{2}. \text{ We have}$$

$$\int x \, \ln(x) \, dx = \frac{x^2}{2} \ln(x) - \int \frac{x^2}{2} \left(\frac{1}{x} \right) dx$$

$$= \frac{x^2}{2} \ln(x) - \frac{1}{2} \int x \, dx$$

$$= \frac{x^2}{2} \ln(x) - \left(\frac{1}{2} \right) \left(\frac{x^2}{2} \right) + C$$

$$= \frac{x^2}{2} \ln(x) - \frac{x^2}{4} + C \text{ for all real values of } x > 0. \, \textbf{(Integration by parts)}$$

14. $\int y e^{3y} \, dy$ for all real values of y.

Let $u = y$ and $dv = e^{3y} \, dy$. Then,

$$du = dy \text{ and } v = \frac{1}{3} e^{3y}. \text{ We have}$$

$$\int y e^{3y} \, dy = \frac{1}{3} \, y e^{3y} - \frac{1}{3} \int e^{3y} \, dy$$

$$= \frac{1}{3} \, y e^{3y} - \left(\frac{1}{3}\right)\left(\frac{1}{3}\right) e^{3y} + C$$

$$= \frac{1}{3} \, y e^{3y} - \frac{1}{9} \, e^{3y} + C \text{ for all real values of y. } \textbf{(Integration by parts)}$$

15. $\int \dfrac{\ln(t)}{t} \, dt$ for all real values of $t > 0$.

Let $u = \ln(t)$ and $dv = \dfrac{1}{t} \, dt$. Then,

$du = \dfrac{1}{t} \, dt$ and $v = \ln(t)$. We have

$$\int \frac{\ln(t)}{t} \, dt = [\ln(t)]^2 - \int \frac{\ln(t)}{t} \, dt$$

We now have the same integral on both sides of the equation, but with different coefficients. Hence, combining them, we have

$$2 \int \frac{\ln(t)}{t} \, dt = [\ln(t)]^2$$

$$\int \frac{\ln(t)}{t} \, dt = \frac{1}{2} \, [\ln(t)]^2 + C \text{ for all real values of } t > 0. \ \ \textbf{(Integration by parts)}$$

16. $\int t e^{-t} \, dt$ for all real values of t.

Let $u = t$ and $dv = e^{-t} \, dt$. Then,

du = dt and $v = -e^{-t}$. We have

$\int t e^{-t} \, dt = -t e^{-t} + \int e^{-t} \, dt$

$$= -t e^{-t} - e^{-t} + C \text{ for all real values of t. } \textbf{(Integration by parts)}$$

17. $\int \dfrac{p}{e^{4p}} \, dp$ for all real values of p.

Let $u = p$ and $dv = e^{-4p} \, dp$. Then,

du = dp and $v = \dfrac{-1}{4} \, e^{-4p}$. We have

$$\int \frac{p}{e^{4p}} \, dp = \frac{-1}{4} \, p e^{-4p} + \frac{1}{4} \int e^{-4p} \, dp$$

$$= \frac{-1}{4} \, p e^{-4p} + \left(\frac{1}{4}\right)\left(\frac{-1}{4}\right) e^{-4p} + C$$

$$= \frac{-1}{4} \, p e^{-4p} - \frac{1}{16} \, e^{-4p} + C \text{ for all real values of p. } \textbf{(Integration by parts)}$$

18. $\int \ln(p^2) \, dp$ for all real values of $p \neq 0$.

Let $u = \ln(p^2)$ and $dv = dp$. Then,

$$du = \frac{1}{p^2} (2p) \, dp \text{ and } v = p$$

$$= \frac{2}{p} \, dp$$

We have

$$\int \ln(p^2) \, dp = p \ln(p^2) - \int \frac{2}{p} (p) \, dp$$

$$= p \ln(p^2) - 2 \int dp$$

$$= p \ln(p^2) - 2p + C \text{ for all real values of } p \neq 0. \text{ (Integration by parts)}$$

19. $\int x^2 e^x \, dx$ for all real values of x.

Let $u = x^2$ and $dv = e^x \, dx$. Then,

$du = 2x \, dx$ and $v = e^x$. We have

$$\int x^2 e^x \, dx = x^2 e^x - 2\int xe^x \, dx.$$

We now integrate by parts again. We have

$$\int x^2 e^x \, dx = x^2 e^x - 2\int xe^x \, dx.$$

Let $u = x$ and $dv = e^x \, dx$. Then,

$du = dx$ and $v = e^x$. We have

$$\int x^2 e^x \, dx = x^2 e^x - 2[xe^x - \int e^x \, dx]$$

$$= x^2 e^x - 2xe^x + 2e^x + C \text{ for all real values of } x. \text{ (Integration by parts)}$$

20. $\int x \ln(1 + x) \, dx$ for all real values of $x > -1$.

Let $u = \ln(1 + x)$ and $dv = x \, dx$. Then,

$$du = \frac{1}{1 + x} \, dx \text{ and } v = \frac{x^2}{2}. \text{ We have}$$

$$\int x \ln(1 + x) \, dx = \frac{x^2}{2} \ln(1 + x) - \frac{1}{2}\int \frac{x^2}{1 + x} \, dx$$

$$= \frac{x^2}{2} \ln(1 + x) - \frac{1}{2}\int (x - 1 + \frac{1}{1 + x}) \, dx \quad \text{(Divide)}$$

$$= \frac{x^2}{2} \ln(1 + x) - \frac{1}{2}\left[\frac{x^2}{2} - x + \ln(1 + x)\right] + C$$

$$= \frac{x^2}{2} \ln(1 + x) - \frac{x^2}{4} + \frac{x}{2} - \frac{\ln(1 + x)}{2} + C \text{ for all real values of } x > -1. \text{ (Integration by parts)}$$

21. $\int y^3 \ln(y) \, dy$ for all real values of $y > 0$.

Let $u = \ln(y)$ and $dv = y^3 \, dy$. Then,

$$du = \frac{1}{y} \, dy \text{ and } v = \frac{y^4}{4}. \text{ We have}$$

$$\int y^3 \ln(y) \, dy = \frac{y^4}{4} \ln(y) - \frac{1}{4}\int y^3 \, dy$$

$$= \frac{y^4}{4} \ln(y) - \left(\frac{1}{4}\right)\left(\frac{y^4}{4}\right) + C$$

$$= \frac{y^4}{4} \ln(y) - \frac{y^4}{16} + C \text{ for all real values of } y > 0. \textbf{ (Integration by parts)}$$

22. $\int [\ln(x)]^2 \, dx$ for all real values of $x > 0$.

Let $u = [\ln(x)]^2$ and $dv = dx$. Then,

$$du = 2 \ln(x) \left(\frac{1}{x}\right) dx \text{ and } v = x. \text{ We have}$$

$$\int [\ln(x)]^2 \, dx = x \, [\ln(x)]^2 - 2 \int \ln(x) \, dx.$$

Let $u = \ln(x)$ and $dv = dx$. Then,

$$du = \frac{1}{x} \, dx \text{ and } v = x. \text{ We have}$$

$$\int [\ln(x)]^2 \, dx = x \, [\ln(x)]^2 - 2 \, [\, x \ln(x) - \int dx \,]$$

$$= x \, [\ln(x)]^2 - 2x \ln(x) + 2x + C \text{ for all real values of } x > 0. \textbf{ (Integration by parts)}$$

23. $\int x^2 e^{-x} \, dx$ for all real values of x.

Let $u = x^2$ and $dv = e^{-x} \, dx$. Then,

$$du = 2x \, dx \text{ and } v = -e^{-x}. \text{ We have}$$

$$\int x^2 e^{-x} \, dx = -x^2 e^{-x} + 2 \int x e^{-x} \, dx.$$

Let $u = x$ and $dv = e^{-x} \, dx$. Then,

$$du = dx \text{ and } v = -e^{-x}. \text{ We have}$$

$$\int x^2 e^{-x} \, dx = -x^2 e^{-x} + 2[-x e^{-x} + \int e^{-x} \, dx]$$

$$= -x^2 e^{-x} - 2x e^{-x} + 2 \int e^{-x} \, dx$$

$$= -x^2 e^{-x} - 2x e^{-x} - 2e^{-x} + C \text{ for all real values of } x. \textbf{ (Integration by parts)}$$

24. $f(x) = x; \, [0, 1]$

Average value of f on $[0, 1] = \dfrac{1}{1 - 0} \int_0^1 x \, dx = \int_0^1 x \, dx = \left[\dfrac{x^2}{2}\right]_0^1 = \dfrac{1}{2} - 0 = \dfrac{1}{2}$. **(Average value of function)**

25. $f(x) = x^2 + 1; \, [-1, 2]$

Average value of f on $[-1, 2] = \dfrac{1}{2 - (-1)} \int_{-1}^2 (x^2 + 1) \, dx$

$$= \frac{1}{3} \int_{-1}^2 (x^2 + 1) \, dx = \frac{1}{3}\left[\frac{x^3}{3} + x\right]_{-1}^2 = \frac{1}{3}\left[\left(\frac{8}{3} + 2\right) - \left(\frac{-1}{3} - 1\right)\right]$$

$$= \frac{1}{3}\left[\frac{8}{3} + 2 + \frac{1}{3} + 1\right] = \frac{1}{3}(6) = 2. \textbf{ (Average value of function)}$$

26. $g(x) = x^3 + 2; \, [-2, 2]$

Average value of g on $[-2, 2] = \dfrac{1}{2 - (-2)} \int_{-2}^2 (x^3 + 2) \, dx$

$$= \frac{1}{4} \int_{-2}^2 (x^3 + 2) \, dx = \frac{1}{4}\left[\frac{x^4}{4} + 2x\right]_{-2}^2 = \frac{1}{4}[(4 + 4) - (4 - 4)]$$

$$= \frac{1}{4}(8 - 0) = 2.\ \textbf{(Average value of function)}$$

27. $h(x) = x\sqrt{4 - x^2}$; $[0, 2]$

Average value of h on $[0, 2] = \dfrac{1}{2 - 0}\displaystyle\int_0^2 x\ \sqrt{4 - x^2}\ dx$

$$= \frac{1}{2}\int_0^2 x\sqrt{4 - x^2}\ dx = \left(\frac{1}{2}\right)\left(\frac{-1}{2}\right)\int_0^2 x\sqrt{4 - x^2}\ (-2x)\ dx$$

$$= \left(\frac{1}{2}\right)\left(\frac{-1}{2}\right)\left[\frac{(4 - x^2)^{3/2}}{3/2}\right]_0^2 = \frac{-1}{6}\left[(4 - x^2)^{3/2}\right]_0^2$$

$$= \frac{-1}{6}(0^{3/2} - 4^{3/2}) = \frac{-1}{6}(0 - 8) = \frac{4}{3}.\ \textbf{(Average value of function)}$$

28. $p(x) = \dfrac{x^2 + 3}{x^2}$; $[1, 4]$

Average value of p on $[1, 4] = \dfrac{1}{4 - 1}\displaystyle\int_1^4 \frac{x^2 + 3}{x^2}\ dx$

$$= \frac{1}{3}\int_1^4 \frac{x^2 + 3}{x^2}\ dx = \frac{1}{3}\int_1^4 \left(1 + \frac{3}{x^2}\right)dx$$

$$= \frac{1}{3}\left[x + 3\left(\frac{x^{-1}}{-1}\right)\right]_1^4 = \frac{1}{3}\left[x - \frac{3}{x}\right]_1^4 = \frac{1}{3}\left[\left(4 - \frac{3}{4}\right) - (1 - 3)\right]_1^4$$

$$= \frac{1}{3}\left[\frac{13}{4} + 2\right] = \left(\frac{1}{3}\right)\left(\frac{21}{4}\right) = \frac{7}{4}.\ \textbf{(Average value of function)}$$

29. $t(x) = \sqrt{x}\ (x - 1)$; $[0, 1]$

Average value of t on $[0, 1] = \dfrac{1}{1 - 0}\displaystyle\int_0^1 \sqrt{x}\ (x - 1)\ dx$

$$= \int_0^1 (x^{3/2} - x^{1/2})\ dx = \left[\frac{x^{5/2}}{5/2} - \frac{x^{3/2}}{3/2}\right]_0^1 = \left[\frac{2}{5} - \frac{2}{3}\right] - (0 - 0)$$

$$= \frac{-4}{15}\ \textbf{(Average value of function)}$$

30. $t(x) = x^3 - 2x$; $[-4, -1]$

Average value of t on $[-4, -1] = \dfrac{1}{-1 - (-4)}\displaystyle\int_{-4}^{-1}(x^3 - 2x)\ dx$

$$= \frac{1}{3}\int_{-4}^{-1}(x^3 - 2x)\ dx = \frac{1}{3}\left[\frac{x^4}{4} - x^2\right]_{-4}^{-1}$$

$$= \frac{1}{3}\left[\left(\frac{1}{4} - 1\right) - (64 - 16)\right] = \frac{1}{3}\left(\frac{-3}{4} - 48\right) = \left(\frac{1}{3}\right)\left(\frac{-195}{4}\right)$$

$$= \frac{-65}{4}.\ \textbf{(Average value of function)}$$

Grade Yourself

Circle the numbers of the questions you missed, then fill in the total incorrect for each topic. If you answered more than three questions incorrectly, you need to focus on that topic. (If a topic has less than three questions and you had at least one wrong, we suggest you study that topic also. Read your textbook, a review book, or ask your teacher for help.)

Subject: Techniques of Integration

Topic	Question Numbers	Number Incorrect
Integration by substitution	1, 2, 3, 4, 5, 6, 7, 8, 9, 10, 11	
Integration by parts	12, 13, 14, 15, 16, 17, 18, 19, 20, 21, 22, 23	
Average value of function	24, 25, 26, 27, 28, 29, 30	